Edward Step

The Romance of Wild Flowers

A Companion to the British Flora

Edward Step

The Romance of Wild Flowers
A Companion to the British Flora

ISBN/EAN: 9783337067250

Printed in Europe, USA, Canada, Australia, Japan

Cover: Foto ©berggeist007 / pixelio.de

More available books at **www.hansebooks.com**

THE ROMANCE

OF

WILD FLOWERS

A COMPANION TO THE

BRITISH FLORA

BY

EDWARD STEP, F.L.S.
AUTHOR OF
"FAVOURITE FLOWERS OF GARDEN AND GREENHOUSE"
"WAYSIDE AND WOODLAND BLOSSOMS" ETC.

WITH UPWARDS OF
TWO HUNDRED ORIGINAL ILLUSTRATIONS

LONDON
FREDERICK WARNE & CO.
AND NEW YORK
1899

[All Rights Reserved]

PREFACE

THE present work is not intended for botanists, but for unscientific flower-lovers. It makes no pretence to enabling the reader to identify the plants noticed in his rambles; but, having identified them by other aids, the following pages, it is hoped, will invest them with greater interest by calling his attention to those details of their structure or behaviour which suggest the term romance.

Much of what is said in respect of the origin of certain forms is obviously mere conjecture, but the author believes that his presentation of the facts upon which these conjectures are based, is such that the reader will not be misled into attaching a higher value to the latter than they merit. Neither does the writer claim originality for these attempts to indicate the probable significance of organs; on the contrary, he would take this opportunity to acknowledge his many obligations to Darwin, Grant Allen, Henslow, Lubbock, Müller, Ogle, and others, who have in recent years done so much to invest our flowering plants with a romantic charm of which the botanist of a generation or two back never dreamed.

The plan of the work is to consider the principal plant families in their proper sequence as regarded by scientific botanists, but as there are some small differences in the systems of living authorities, the author has thought it wise to keep the present work in harmony with his former flower-books, by following the arrangement of Sir J. D. Hooker's admirable *Student's Flora of the British Islands*.

The illustrations in the text have been specially drawn for this work, and for their execution the author has been principally indebted to the pen of his daughter. The headpieces by Theo. Carreras are also original; and the tinted plates are from photos taken by the author from plants in their natural habitats, and in some cases under very difficult conditions.

CONTENTS

PAGE

I. INTRODUCTORY

Former Notions respecting Plant-Life—Linnean Axiom—The new Botany—Senses of Plants—The *raison d'être* of Form, Colour, and Perfume—Plant Intelligence . 17

II. ROSES AND APPLES

The Garden Rose a Monstrosity—Wild Roses—Structure of Rose-Trees and their Flowers—Fertilisation—"Hips"—Seeds: their Dispersion—Birds and Fruits—Apples—Cross-Fertilisation—Crab-Apples—Wild Pear—Medlar—Rowan—Hawthorn—Arrangement and Forms of Leaves—Cinquefoils—Strawberry—Origin of the Strawberry—Herb Bennet: its Hooked Seeds—Agrimony—Brambles: their Varieties—How Species arise—Blackthorn and Sloe—Plums—Cherry—Lady's Mantle—Dropwort and Meadow-Sweet—Salad Burnet 21

III. BUTTERCUPS AND COLUMBINES

Lesser Celandine—Thrifty Plants—Origin of Tuberous Roots—Buttercups—The Significance of Leaf-Form—The Purpose of Acrid Juices—Spearworts and Goldielocks—Globe-Flower and Marsh Marigold—Anemone and Pasque-flower—Meadow Rues—Forsaken by Insects—Wild Clematis—How it rises in the World—Mousetail: its Economy—Columbines and Bees—Fraudulent Honey-Seekers—Larkspur and Monkshood—Highly Specialised Flowers run Risks 57

IV. POPPIES

Plants with Milky Sap—Horned Poppy—Greater Celandine—Welsh Poppy—Red Poppies—Their Common Origin—Poppies Honeyless 78

V. WALLFLOWER AND CABBAGE

Sweetness and Rankness Allied—Wallflower—An Improvised Tube—Stocks—The Significance of Round and Lobed Stigmas—Transported Water-Cress—Anti-Scorbutics—Lady's Smock—Hedge-Mustards—Wild Cabbage and its Cultivated Descendants—The Origin of Garden "Greens"—Shepherd's Purse well filled—Floral Aristocracy and Labour—Hangers-on of Industry—Candytuft and Woad . 84

VI. VIOLETS AND PANSIES

Violet's conspicuous Flowers rarely produce Seeds—Honey Guides—Modest and Patient Violets—A Resourceful Plant—Flowers that never Open—Vegetable Sharp-Shooters—Wild Pansy—Its Skull-like Stigma and abundant Seeds . 95

VII. PINKS AND CHICKWEED

Red Campion: a Day Flower—White Campion: an Evening Flower—Their Common Origin—Ragged Robin—Bladder Campion and its Insect Enemies—Sea Campion—Maiden Pink—Corn-Cockle—The Mouse-Ears—Chickweeds and Stitchworts—Sandworts—Catchflies and their Digestive Powers 103

VIII. MALLOWS

Marsh-Mallow—Common Mallow—Dwarf Mallow—Differences in the Flowers and the Methods of Fertilisation—How Honey-Stealers are excluded—Tree Mallow . . 118

IX. GERANIUMS

Mahomet and the Mallow—Wood Crane's-bill—Sprengel's Researches—Meadow Crane's-bill—Mountain Crane's-bill—Dove's-foot—Small-flowered Geranium—Round-leaved Geranium—Herb Robert—Shining Crane's-bill—Remarkable Seed-shooting Mechanism—Stork's-bills and their Seed-burying—Wood Sorrel and its Sensitive Leaves—More Never-opening Flowers—Touch-Me-Not and Jewel-Weed—Spring-gun Seed-vessels 123

X. PEAS AND CLOVER

Plants rich in Nitrogen—How they obtain it—Nitrifying Bacteria—Meadow Vetchling—Structure of Pea-Flowers adapted to the Comfort of the Bee—Pea-pods—Illicit Honey-drinkers—Yellow Vetchling—Grass Vetch—Narrow-leaved Everlasting Pea—Intelligent Bees—Bitter Vetch—Wood Vetch—Sain-foin—Horse-shoe Vetch—Bird's-foot Trefoil—Bird's Foot—Kidney Vetch—Rest Harrow—Dyer's Greenweed—Furze—Broom: Strange Behaviour of its Pistil—Lucerne—Melilot—Clovers—Cats, Mice, Bees, and Clover—Subterranean Trefoil—How it buries its Pods . . 137

XI. SUNDEWS

Round-leaved Sundew — A Carnivorous Plant — Fine Sense of the Leaves—Darwin's Experiments—Long-leaved Sundews 158

XII. THE PARSLEY AND CARROT FAMILY

Cow-Parsnip—Umbel-bearers—Vegetable Scent-Bottles—Marsh Pennywort—Hare's-ear—Sea-Holly—Sanicle—Poisonous Umbellifers—Parsley—Its Native Place unknown—Parsnip and Carrot evolved by Cultivation—Umbellifers chiefly fertilised by Flies and Beetles . . . 164

XIII. WOODRUFF AND GOOSEGRASS

Field Madder—Woodruff: how its Seeds are dispersed—Madder: its Fruits sought by Birds—Bedstraws—Crosswort—Hedge-Bedstraw—Field Bedstraw's Climbing-Hooks: its Seeds distributed with the Grain—Goose-Grass: its Hook-covered Stems and Fruits 171

XIV. DAISIES AND THISTLES

Composite Flowers the most numerous and most widely distributed — How they originated — Altruism among Flowers—Guelder-Rose—Honeysuckle—Hemp Agrimony—Structure of the Florets — How fertilised — Daisy—Advanced Co-operative Idea—The Daisy's Advertising Department—Hints as to Evolution—Flea-banes—Chamomiles—Chrysanthemums—Pollen Brushes—Yarrow and Sneezewort—Reasons for Bitterness in Plants—Tansy—Wormwood—Coltsfoot—Butterbur—Seed Parachutes—Ragworts—Groundsel—Fœtid Odour—Burdock—Hooked

Bracts — Carline Thistle — Knapweeds — Cornflower — Thistles—Vegetable Spider-Webs—Goldfinches and Thistle-down Plume-Thistles—Creeping Plume-Thistle laughs at the Sickle—Thistles as Food—Holy-Thistle—The *Raison d'être* of Thistle-Spines—Chicory—Dandelion—How its Seed is planted—Business Hours of Flowers—Teasel—Offence and Defence—Turning the Tables on the Honey-Thieves—Bovril for Plants 175

XV. HAREBELLS

Round-leaved Bell-Flower — Developmental Clues — Nettle-leaved Bell-Flower — Spreading Bell-Flower — Rampion: protected from Beasts, but not from Man—Venus' Looking-Glass—Ivy-leaved Bell-Flower—Sheep-Bit—Round-headed Rampion—Lobelia 202

XVI. BILBERRY AND HEATHER

"Hurts"—Singular Stamens of the Bilberry Flower—Bog Whortleberry—Cranberry—Strawberry-Tree or Arbutus—Bearberry—Purple Heath—Cross-leaved Heath—Fringed Heath—Cornish Heath—Irish Heath—Atlantis and Lyonesse—Heather—St. Dabeoc's Heath—Menziesia and Trailing Azalea—Survivals from the Great Ice Age . . 208

XVII. PRIMROSE AND PIMPERNEL

Primrose specially adapted for Insect Fertilisation—How its Tube has been formed—Long-styled and Short-styled Forms—Darwin's Discovery—Cowslip—Scottish Primrose—Bird's-eye—Artful Bees again—Yellow Loosestrife—Creeping Jenny produces no Seeds in Britain—Woodland Loosestrife—Sea Milkwort—Pimpernel: its Early-Closing Habit—Bog Pimpernel—Water Violet . . . 218

XVIII. GENTIAN AND BOGBEAN

Gentians Alpine Plants—Field Gentian fertilised by Humble-Bees—Felwort—Marsh Gentian—Spring Gentian—Small Alpine Gentian—The Ancestral Gentian Yellow and Open—Evolution of Form and Colour—Yellow-Wort—Centaury—Bogbean 229

XIX. BUGLOSS AND SCORPION GRASS

Viper's Bugloss — Borage — Prickly Comfrey — Common Bugloss — Forget-me-Nots or Scorpion Grasses — Hooked Fruits carried by Man and Beast — Clear Proof of the Reasons for Hairy Stems and Fruits 235

XX. FOXGLOVE AND TOADFLAX

A Grand Flower Show organised by Humble-Bees — Methodical Bees — No Small Bees need apply — Unusual Arrangement of Organs — Mulleins — Speedwells fertilised by Flies — The Corolla thrown off — Brooklime — Spiked Speedwell — Figwort and Wasps — The Wasp's Methods different from those of the Bee — Water Figwort — Monkey-Flower and Sensitive Stigmas — Money-Wort — Root-Parasites — Cow-wheats — Yellow-Rattle — Bartsias — Eyebright — Lousewort — Only partially Dishonest — "The Rival Broom-sellers" — Broomrapes and Toothwort — Toadflax — The Intelligent Ivy-leaved Toadflax — Yellow Toadflax . 242

XXI. BUTTERWORT

Poor Soil develops Murderous Instincts — Butterwort — Bladderworts — Irritable Stigmas — Submerged Traps for Insects and Crustaceans 267

XXII. MINT AND THYME

Aromatic Plants — Mints — Protected by Strong Odours — Gipsy-Wort — Marjoram — Wild Thyme — Meadow-Sage — A Forgotten Record — Clary — Dishonest Butterflies — Ground Ivy — Self-Heal — Woundwort — Hemp Nettle — Dead Nettles not related to Stinging Nettles — Yellow Archangel — Henbit — Black Horehound — Wood-Sage — Bugle — Ground Pine 273

XXIII. SPURGES AND NETTLES

Sun Spurge — Irish Spurge — A Poisonous Family — Box — Mercury — Elms: how their Seeds are dispersed — Stinging Nettles fertilised by the Wind — Mechanism of the Nettle's Sting — Why the Nettle stings — Hop, How it climbs — An Exploring Stem — Darwin's Illustration — Forest Trees . 287

XXIV. ORCHIDS

Many Native Species—Structure of Orchid Flowers—An Interesting Experiment—Suicidal Flies—Man Orchis—Bee Orchis—Fly Orchis—Musk Orchis—Fragrant Orchis—Butterfly Orchis—Bog Orchis—Coral-Root and Bird's-Nest—Probable History—Tway-Blade—Lady's Tresses—Helleborines—Lady's Slipper 297

XXV. FLAG AND CROCUS

Blue-eyed Grass—Romulea—Autumnal Crocus—Vernal Crocus—Two Modes of Propagation—Young Corms plant themselves—Yellow Iris—Its Puzzling Flowers—Fly-Flowers and Bee-Flowers—Gladdon or Roast-beef Plant . 311

XXVI. DAFFODIL AND SNOWDROP

Links between Flags and Lilies—Underground Treasuries—Lent-Lily—Snowdrop—A Business Day of Six Hours—More Thriftiness—Snowflakes 318

XXVII. LILIES AND ONIONS

Odour and Malodour related—Martagon Lily—An Ingenious Plan for economising Honey—Nature of Lily-Bulbs—How Lilies travel—Alexander Pope in need of Revision—The Founder of the Lily Family—Asparagus—Butcher's Broom: Why it is Leafless—Lily of the Valley—A Floral Fraud—Solomon's Seal—Broad-leaved Garlic—Wild Leek—Grape Hyacinth—Vernal Squill—Autumnal Squill—Blue Bell—Star of Bethlehem—Fritillary or Snake's-head—Wild Tulip—Alpine Tulip: a probable Survivor of the Great Ice Age—Meadow Saffron—Bog Asphodel—Scottish Asphodel—Herb-Paris mimicks Carrion and deludes Flies 323

XXVIII. RUSHES AND REEDS

Degenerate Lilies—Plants that have "come down in the World"—Wood Rush Wind-fertilised—Rushes—Bulrush—Sedges—Cuckoo-Pint—Systematic Deception—False Pretences followed by Illegal Detention—From Tavern to Tavern—Sweet Flag 339

XXIX. GRASSES

Structure and Characteristics of Grasses—The Great Reed—The Marraw or Sea Reed—Onion Couch—Spikelets and Glumes—Grasses all Wind-fertilised—Feathery Stigmas . 346

LIST OF ILLUSTRATIONS

PAGE PLATES

1. Samphire	*Frontispiece*	16. Honeysuckle *Facing page*	177
2. Mouse-ear Hawk-weed	*Facing page* 17	17. Hemp Agrimony ,,	178
		18. Flea-bane ,,	182
3. Meadow Sweet ,,	54	19. Ragwort ,,	187
4. Wood Anemone ,,	70	20. Carline Thistle ,,	188
5. Common Poppy ,,	82	21. Hard-heads ,,	191
6. Red Campion ,,	104	22. Spear Plume-thistle ,,	194
7. Sea Campion ,,	108	23. Chicory ,,	196
8. Stitchwort ,,	112	24. Teasel ,,	201
9. Narrow-leaved Ever-lasting Pea ,,	145	25. Sheep's-bit ,,	206
		26. Purple Heath ,,	214
10. Rest Harrow ,,	148	27. Primrose ,,	220
11. Hare's-foot Trefoil ,,	156	28. Comfrey ,,	238
12. Cow Parsnip ,,	164	29. Foxglove ,,	242
13. Sea Holly ,,	167	30. Ivy-leaved Toadflax ,,	263
14. Wild Carrot ,,	169	31. Yellow Toadflax ,,	265
15. Stinking Mayweed ,,	175	32. Yellow Flag ,,	315

TEXT ENGRAVINGS

	PAGE		PAGE
INTRODUCTORY	17	Opening bud	26
ROSES AND APPLES	21	Stamen	26
Wild Rose	22	Pollen-grain	27
Rose-leaf	23	Pistil	27
Rose-bud	24	Section of Rose	28
Moss Rose-bud	24	"Hips"	29

	PAGE
Rose Nutlet	30
Section of Hip	30
"Hips"	31
Section of Nutlet	31
Apple-flower	32
Apples	34
Section of Apple-flower	34
Section of Apple	35
"Haws"	36
Cinquefoil	39
Strawberry-leaf	41
Wild Strawberry	42
Wood Avens	44
Agrimony	45
Cloudberry	47
Blackberry	48
Sloe	50
Cherry Blossom	51
Cherries	52
Dropwort	54
Dropwort Flower	54
BUTTERCUPS AND COLUMBINES	57
Lesser Celandine	58
Lesser Celandine, Petal with Nectary	58
Lesser Celandine's fleshy roots	59
Section of Celandine	60
Bulbous Buttercup	65
Globe-flower	66
Petal of Globe-flower	67
Wood Anemone	68
Pasque-flower, Stamen and Nectary	70
Seed-head of Clematis	71
Monkshood	75
Section of Monkshood	76
Anthers of Monkshood: those curled back have shed their pollen	76
Anthers all curled back out of way of ripe stigmas	76
POPPIES	78
Horned Poppy	79
Long Prickly-headed Poppy	81
WALLFLOWER AND CABBAGE	84
Wallflower	85
Stamens and Style of Wallflower	87
Seed-vessel of Wallflower	87
Shepherd's Purse	92
VIOLETS AND PANSIES	95
Section of Violet	96
Violet Capsule, full of seeds	99
Violet Capsule, seeds discharged	99
Ovary and Stigma of Wild Pansy	100
Section of Pansy Ovary	101
PINKS AND CHICKWEED	103
Sea Campion	107
Sea Campion, Male flower	108
Sea Campion, Female flower	108
Sea Campion, Female flower enlarged	108
Sea Campion, Petal and Stamen	108
Maiden Pink	109
Chickweed Flower	114
MALLOWS	118
Stamens of Common Mallow	120
Stigmas of Common Mallow	121

	PAGE		PAGE
GERANIUMS	. 123	PARSLEY AND CARROT	. 164
Geranium scattering seeds	. 129	Fruit of Cow Parsnip	. 166
		WOODRUFF AND GOOSE-GRASS	. 171
Herb-Robert discharging its seed	. 129	Woodruff	. 172
Stork's-bill seed	. 131	DAISIES AND THISTLES	. 175
Wood Sorrel	. 132	Hemp Agrimony	. 178
Section of Impatiens Flower	. 135	Portion of Daisy-head showing arrangement of florets	. 179
Seed-vessel of Impatiens	135	Ray-floret of Daisy	. 180
Capsule valves coiling up and discharging seeds	. 136	Disk-floret of Daisy	. 181
		Dandelion floret	. 197
PEAS AND CLOVER	. 137	Dandelion seed and pappus	198
Bacteria nodules on roots of White Clover	. 138	Field Scabious and Fruit	200
		Teasel Flower	. 201
Meadow Vetchling	. 140	HAREBELLS	. 202
Parts of a Leguminous Flower	. 141	Harebell, 1st condition	. 204
		Harebell, 2nd condition	. 204
Stamens and Pistil of Pea-flower	. 142	BILBERRY AND HEATHER	. 208
		Bilberry Flower	. 209
Pea-pod opened from back	. 142	Bilberry stamen	. 210
		Bilberry	. 210
Grass Vetch	. 143	PRIMROSE AND PIMPERNEL	218
Seedling Furze	. 149	Primrose, Long-styled	. 219
Broom	. 150	Primrose, Short-styled	. 219
Broom, 1st condition	. 150	Glaux opening	. 224
Broom, 2nd condition	. 151	Glaux, Section of flower	225
White Clover, 1st stage	. 154	Glaux, Natural position	. 225
White Clover, Fertilised flowers turned down	. 154	Pimpernel	. 226
		Water Violet, Long-styled	. 227
White Clover, all Fertilised	. 154	Water Violet, Short-styled	. 227
White Clover, side view	155		
White Clover, Upright flower	. 156	GENTIAN AND BOGBEAN	. 229
		Field Gentian	. 230
White Clover, Calyx and Standard removed	. 156	BUGLOSS AND SCORPION GRASS	. 235
SUNDEWS	. 158	Viper's Bugloss	. 236
Round-leaved Sundew-leaf	. 159	Fruiting calyx of Forget-me-not	. 240

	PAGE
FOXGLOVE AND TOADFLAX	242
Three conditions of the Foxglove	244
Germander Speedwell	248
Bursting bud of Speedwell	250
Flattened calyx of Speedwell	250
Ivy-leaved Toadflax	263
Yellow Toadflax	265
BUTTERWORT	267
Section of Butterwort	268
Bladderwort's trap	270
MINT AND THYME	273
Meadow Sage	277
Stamen of Meadow Sage	277
White Dead-nettle	281
Three stages of Wood Sage flower	284
SPURGES AND NETTLES	287
Flowers of Petty Spurge	289
Nettle sting	292
Nettle, Male flower, 1st state	293
Nettle, Male flower, 2nd state	293
Nettle, Female flower	293
Sallow-bloom	296
1. Female catkin	296
2. Male ,,	296
3. Pistillate flower	296
4. Staminate ,,	296
ORCHIDS	297
Early Purple Orchis	298
Early Purple Orchis	299
Pol. Pollinia	299
Ro. Rostellum	299
St. Stigma	299
Early Purple Orchis, side view	300
Fly Orchis	303
Butterfly Orchis	304
Lady's Slipper	309
Section of Lady's Slipper	310
FLAG AND CROCUS	311
Vernal Crocus	313
DAFFODIL AND SNOWDROP	318
Snowdrop	320
Section of Snowdrop	320
Snowdrop's stamen	321
LILIES AND ONIONS	323
Petal and Nectary of Martagon Lily	324
Asparagus, Female	328
Asparagus, Male	328
Butcher's Broom	329
Lily of the Valley	330
Herb Paris	338
RUSHES AND REEDS	339
Wood-rush	340
Cuckoo-pint	341
Spadix of Cuckoo-pint	342
GRASSES	346
Grass spikelet	348
Grass-flower	349

Mouse-ear Hawk-weed.

INTRODUCTORY

IT may be fairly claimed that during the last half-century our prevailing notions respecting plant-life have been greatly modified, and concerning flowering plants have been entirely changed. Fifty years ago there could be found very few botanists who were not satisfied with the generalisations crystallized in the Linnean axiom—

> Stones grow,
> Vegetables grow and *live*,
> Animals grow, live, and *feel*."

There, in less than a dozen words, was a handy and easily remembered formula, serving as a kind of touchstone which, when applied to any doubtful organism, would detect whether it were plant or animal. But other times, other methods; the Linnean formula is absolutely obsolete to-day. The increase in bulk of crystals by additions to their surfaces would not be regarded to-day as analogous to the growth of plants and animals. The modern botanist can bring to your notice numberless examples of plant-life which not only exhibit sensitiveness to touch, but also the power of free locomotion. There are plants that catch living animals and put them to death, afterwards digesting and assimilating their victims. There are organisms in the lower ranges of life that are at once so plant-like and so animal-like that botanist and zoologist both claim them as subjects. There are others whose vegetable nature would scarcely have been suspected but for the distinctly modern practice of closely observing and studying the developmental changes of organisms. Plants we now know to be sensitive to light, to atmospheric conditions, and in certain cases to be able to distinguish between different classes of matter regarded from the chemical standpoint.

Then, too, respecting flowers, recent research has enabled us to take a higher view of their importance. The characteristically conceited notion man cherished for ages respecting them was that their *raison d'être* was to please him by their forms, their colours, and their perfumes, and the desert flower that was born to blush unseen " wasted " its fragrance because no man passed that way. We of to-day know that the desert flower and its ancestors were efficiently performing their functions in the machinery of the universe though for a thousand years or more no man had come within sight. Were flowers capable of doing reverence, it is not man they would acknowledge as lord, but the bee or butterfly. For these humble creatures—ignorant no doubt of their importance—does the flower develop its most brilliant tints; for these it distils its choicest nectar, and flings abroad the sweetest odours; for these it actually modifies its form and arranges its most essential organs; for these are its hours of expansion and of folding its petals fixed, and probably its season of appearance also.

There is much greater similarity between animals and plants in their structure and their vital processes than the non-scientific public is disposed to recognise, widely though they may differ in the forms and action of their organs. The highest and the lowest plants, like the simplest and most complex of animals, are built up of minute cells, by whose subdivision growth proceeds. There is a circulation of vital fluids, an alimentary process by which the plant is fed, a breathing system by which the inhaled air is decomposed—one gas being retained, another given off, a reproductive process presenting close analogies with that of the animal kingdom, and the employment of various "dodges" to effect certain ends that would do credit to animals with brain-power of a high order.

So far from the sense of feeling being an attribute of animals alone, many plants are most irritable. Professor George Henslow, indeed, holds this irritability chiefly responsible for the peculiar structure of the best known flowers, many of which are obviously modelled with a special view to the visits of certain insects. Then, again, the sensitiveness of the stems and tendrils of climbing plants is of a high order, and the plant's response to the impression given by contact would certainly be regarded as evidence of intelligence if displayed by an animal. It is indeed essential that the reader of the following pages should at starting get rid of any preconceived notions based on natural history works of a generation ago. Until this had been done by modern observers, botany remained, and was like to remain, a matter of counting petals and stamens, using long words of Greek or Latin origin, and knowing a little of the reputed medicinal virtues of each species. To-day the naturalist is more concerned in observing the habits of the plant, especially its attitude towards insects, and the natural relation of one species to another as revealed in the main structural features of the flower, and by comparing the smaller differences which separate species obviously built on the same general plan.

In the present work the author aims at giving a brief account of the general characteristics of the principal families of British flowering plants, in simple language, devoid of technicalities so far as these are not absolutely necessary for clearness. The facts concerning the internal structure of roots, stems and leaves, of cells and tissues, and the chemistry of plant-life will be taken for granted, as these may be obtained from any elementary treatise, and to deal with them here would take up too many of the pages required for a different branch of the subject. The names of the different floral organs will recur on every page, and it is well that the reader should start with a very clear notion of what they imply. We shall endeavour to make this all plain in our first chapter, though this will have the effect of giving an air of "dryness" to our start.

ROSES AND APPLES

IT comes like a shock to the enthusiastic Rose-grower who is no botanist, to tell him that all the lovely blossoms upon which he sets such store are, strictly speaking, monstrosities. He will probably seek to rebut your assertion by pointing to the exquisite forms and delicate shades of colour of his pets, whose beauty is enhanced in many cases by the exhalation of subtle and delicious perfume. "How," he may ask, "can you apply the term 'monstrous' to such beautiful creations as these?" The term certainly appears on the surface to be misapplied, yet, in judging flowers, we are bound to consider not the arbitrary rules of floricultural exhibition judges, or the varying standards of beauty set up by mankind, but the purpose for which the flowers exist in nature, and how far the particular specimens in question are fitted to serve that purpose.

The first and last purpose of the flower is to insure the continuance of the species by the production of good seed; but the most perfect Roses from the Rose-grower's point of view are those that are least fitted to produce seed. For the really perfect Rose we must look not to the beds of the Rose-garden, but to the thick untrimmed hedges, and the bosky wastes where the Wild Roses (*Rosa arvensis* and *Rosa canina*) exhibit their less pretentious beauty. A careful examination of a few of these Wild Roses will help us to understand how and why the *Gloire de Dijon* and the *Marshal Neil* are monstrosities, and it will give us insight into the main features of flower-structure.

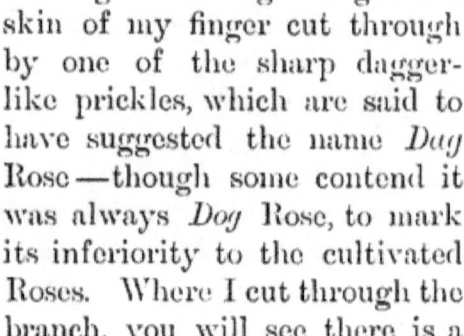

Wild Rose

I cut a long spray from the scrambling plant, and in doing so manage to get the skin of my finger cut through by one of the sharp dagger-like prickles, which are said to have suggested the name *Day* Rose —though some contend it was always *Dog* Rose, to mark its inferiority to the cultivated Roses. Where I cut through the branch, you will see there is a little hard wood which was formed last year; the newer end is soft and green, though tough and stringy when I try to break it. This woody character of the stem denotes a tree or shrub, and as these Rose-plants do not grow tall enough, nor have sufficiently stout trunks to be called trees, they are shrubs. Now, looking along this spray from end to end, we see that, in addition to the hooked prickles, it bears leaves and flowers—very dissimilar things, yet those who have made a very careful study of the matter will tell you that in their origin they are very similar, in fact, identical. The germ or first beginnings of a leaf may be developed into either a leaf or one of the several parts of a flower; whilst the hooks are modified hairs.

The Roses are all more or less climbers, and the strong hooks help them to scramble up between the branches of bushes and hedgerows. If any kind of Rose exhibits a tendency to have only short, straight stems and branches, you will find that the spines are scarcely, if it all, hooked or curved; they may even be reduced to stiff bristles. Should anyone present you with a Rose with a small piece of the stem attached, you could tell pretty well by glancing at the character of the spines whether the plant was a stiff dwarf bush or a climber.

Here is a leaf, broken up into five little *leaflets* which you might be inclined to consider as five separate and complete leaves. The leaflets are all attached to what is really the midrib of the leaf as a whole; and if we were to lay the entire leaf on paper and draw a curved pencil line backwards from the tip of the odd leaflet, so that it just touched the tips of the other leaflets, we should have an outline of ordinary leaf form

Rose-leaf

— like that of an Apple-leaf, for instance. All leaves that are broken up into leaflets in this manner are known as *compound* leaves, and such compound leaves as have the leaflets equally on either side of the midrib are said to be *pinnate* leaves, because this disposition of its parts is similar to the arrangement of the parts of a feather, for which *pinna* is the Latin word. We should call this an unequally pinnate leaf because of the odd leaflet. Now look at the base of the leaf-stalk, and you will notice a pair of wings, so to speak; these are called *stipules*. This is all very tedious and uninteresting, no doubt, but if we wish to understand all about flowers, we must also know the details of structure of the plants that bear the flowers, and perhaps before we have gone very far these details may prove to have an interest of their own. Now let us look at the Dog Rose-flower.

At the top of the flower-stalk is a smooth green egg-shaped knob which can be best observed in this unopened bud, of which it will be well to make a vertical section. It is known as the *receptacle*, and though that strikes us as an appropriate name looking at its hollowed character in the Rose, we shall find in other plants that it is more often flat or convex, sometimes indeed conical. The receptacle, then, is the enlarged head of the flower-stalk from which arise the various floral organs.

Rose-bud

In this Rose there grow up from the rim of the receptacle five rather thick green leaves with ragged edges, which are called *sepals*. The term is derived from a Latin word *sepio*, to hedge around or enclose, and as we see them in the but slightly open bud, the term appears very fit, for they form a hedge or enclosure round the pink Rose-leaves. Collectively these sepals are called the *calyx*, and in many flowers the sepals are so closely united by their edges, that they can only be spoken of in the collective sense. Calyx means cup, and though

Moss Rose-bud

its appropriateness is not so evident in the case of the Rose—where the sepals are mere outgrowths from the edge of the receptacle—in those flowers whose receptacle is flat the sepals do convert it into a cup.

Now we come to the floral organs that attract the eye and give most pleasure—the beautiful flower-leaves or *petals*. This name *petal* comes from the Greek word *petalon*, a leaf or thin plate of metal. If you were to take the petal of a Buttercup, you would say the Greek word was very appropriate, for it looks like a little bit of gold-leaf thinned out under the gold-beater's hammer. One of the most lovely things in nature is that pink Rose-petal, so delicately tinted, so exquisitely soft and silky. Well, there are five of these petals also, and if we wish to speak of the whole five together, as we did in the case of the sepals, we call them the *corolla*. That is another Latin word, and it means *a little crown*. Now look at this half-open bud and say, is not little crown a fine and fitting word?

It may be asked why not say little crown in English and be done with it, and the same with other parts of the flower? There is really no reason why you should not do so if you prefer it, but botanists all the world over have agreed upon all these parts of a plant being named by words having either Greek or Latin origin, so that no matter what may be their own language, they can understand exactly what each means when he uses these words. They form a universal language, which makes it easy to spread knowledge throughout the world, and often the use of such a term saves a great deal of explanatory writing every time one wishes to refer to a particular thing. It is like the international agreement upon the symbols used to denote the notes, compass, and marks of expression in writing music. A Turk or a Russian may compose a piece of music, and an English schoolgirl who knows nothing of the Turkish or the Russian language can play that musical composition as easily as if it had been written by an Englishman. These technical terms to denote the floral organs will be used on every page of this book, and if I were not to use them, I should have constantly to indulge in the repetition of such phrases as the "pink-coloured leaves of the Rose-flower" and "the green leaves of the Rose-flower," to avoid confusion between these and the ordinary foliage of the plant. But even this would be simplicity compared with the task of frequently referring to some of the other organs for which no English synonym can be found.

Opening bud

Let us proceed further in our investigation of the Rose-flower, and get rid of this tedious business of terminology-explaining. In the centre of the receptacle there is a little cushion, and surrounding it a large number of what look like yellowish pins with oval heads, and the points appearing to be stuck into the edges of the central cushion. If I take one of these pins away and lay it upon my dark coat-sleeve you can look at it through this pocket-lens, and you will see that the head is divided into two little pouches, which split along their edges and set free a quantity of what looks like yellow flour. These

Stamen

two pouches together are reckoned as one *anther*, the pin-like stalk is the *filament*, and the entire organ is called a *stamen*. I cannot say that the authors of these names have been so happy as they were in the case of calyx and corolla, for both stamen and filament are from Latin words which alike mean threads, and anther is from a Greek word signifying flowery. Here is an anther splitting along its edge and disclosing the *pollen*, which is Latin for fine flour. So far as appearances go, that also is a good word, but if we were to put a little of the pollen under the microscope we should find it was composed of golden globules, each beautifully carved and decorated after a definite pattern peculiar to the species. I have not got a drawing of Rose-pollen to show you, but here is a figure of a pollen-grain from the Hollyhock.

Pollen-grain

Now we must cut through an open flower and see what is below this cushion in the centre. The cushion is composed of a number of *pistils* with distinct heads, from each of which a stalk goes down into the receptacle, where it ends in an enlargement called a *carpel* or ovary. The upper part is called the style, and ends in the *stigma* or point. If with a fine sharp knife we cut through the ovary, or egg-chamber, we shall find a hollow space, in which is a tiny white seed-egg or *ovule*. This finishes the cutting and pulling to pieces of the Rose-flower. Now let me explain the purpose of these several parts.

Pistil

Let me tell you first what would happen if one of the golden globules of pollen were to be placed upon the stigma: a little protuberance would appear on the under-side of the pollen-grain, and its outer skin would be broken by a shoot from its interior. This shoot would pierce the substance of the stigma and force its way between the minute cells of which the style is built up—one might liken the process to the root of a tree finding its way between the brickwork of a wall, from top to bottom. Throughout its course the pollen-shoot sets up an irritation in the cells it touches, which causes the plant to fill them up with sugar or other nutritive material, upon which the pollen-tube feeds as it goes along. At length it reaches the ovary, and there searches for a seed-egg.

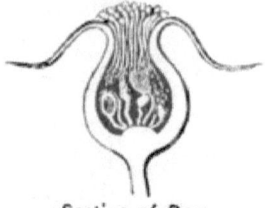

Section of Rose

Each seed-egg is pierced with a minute pore called the *micropyle*, or little gate, and through this little gate the pollen-tube passes, and mingles with the contents, thus giving a stimulus which leads to the formation of an embryo, which develops into a seed capable of producing a Rose-plant like that by which the Rose-flower was produced. Several pollen-grains would perform this office simultaneously and an equal number of seed-eggs would be fertilised. The petals having performed their functions would drop off, and the stamens and stigmas would wither. The receptacle-tube would share the excitement of the embryos, and would grow larger and more fleshy in order to provide room for the enlarging seeds in their carpels, and its outer coat would become bright red and glossy.

To hark back to the freshly expanded Dog Rose: the stamens appear to be all stretching away from the stigmas as though not desiring too close an acquaintance; how then will the pollen-grains reach the stigmas? Many insects are very fond of pollen—some of the wild bees, many flies and small beetles commonly visit the Roses on that account, and as they fly from flower to flower they carry a few pollen-grains clinging to the hairs on their legs and under-sides. These insects almost always fly straight to the centre of the flower and alight on the cushion of stigmas, upon whose sticky surfaces the pollen-grains become detached. Even if they do not bring pollen with them from a previously visited Rose, these creatures are very likely to kick or carry some from the anthers of this particular flower to its own stigmas; but strong efforts are made by many plants to insure cross-fertilisation—that is fertilisation by means of pollen brought from a separate plant of the same species.

Roses do not produce honey as many flowers do, but the petals give out an attractive odour which insects smell, and so are guided to the flower. The flowers that excrete honey have usually fine lines and spots upon their petals indicating the direction in which the honey must be sought. Many examples of these guide-lines will be given in subsequent chapters.

We have reached the point in our Rose-history where the "hips" stand up in their polished ruddy glory on the bushes when autumn has thinned the leaves, and the schoolboy comes along and says, "Here's a splendid lot of hips! Let's have some fun with them!" He gathers a few,

"Hips"

and breaking them open, shakes out the hairy fruits, which are afterwards stealthily slipped down the back of a chum, who is thereby rendered miserable for a whole day. This, however, is not the end the plant had in view in painting and polishing its fruit so well. Its real object was to attract birds. Birds appear to be most partial to red and black among colours; and we shall find that most of the plants which depend upon birds for the dispersal of their seeds adopt one or other of these colours. These Rose-hips are red, but the Sloe, which we shall mention soon, is black, the different hues, no doubt, being appreciated by different races of birds. It will be understood, that if the seeds of the Rose dropped to the ground around the parent bush, there would soon be such keen competition among members of the same species for light, air, and mineral food, that all would suffer, and perhaps die; so it is to the interest alike of the individual and the species that the seeds should be scattered as widely as possible.

Rose Nutlet

The seeds, it will be remembered, are encased in their carpels, or true fruits, which form a kind of nut-shell around them. Some of the smaller birds peck off the soft ripe flesh of the hips, merely scattering the contained fruits, some of which, no doubt, are dispersed further a-field by the investing hairs clinging to the plumage of the bird. But it is probable that birds like the jay, magpie, hawfinch, blackbird, and thrush swallow the hips whole, the fruits

Section of Hip

being rubbed out by the gizzard and passing through the bird's intestines undigested and uninjured. The hairs may expedite this process by setting up internal irritation. It is in this way that the Wild Roses have got themselves planted all along our hedges, our copses, and in every bushy clump upon our heaths.

Hips

Roses grow rapidly because they have not got to wait after every advance of a few inches until they have built up and matured a good thickness of solid wood. By their method of scrambling among the branches of other shrubs and hanging on by the help of their stout hooked prickles, they can add several feet to their stature in one season of growth. These prickles are nothing more than a development of the protective hairs with which many plants are furnished.

The difference between Wild and Garden Roses consists in the fact that the stamens of the Wild Rose have been converted into petals. If we cut down through the centre of a Cabbage Rose, say, we shall find only two or three stamens and some very narrow petals in addition to a large number of fully-developed petals. I do not mean to assert that any of these petals were ever stamens in this particular specimen, but that in the ancestors of the Cabbage

Section of Nutlet

Rose — that is before it became anything like a cabbage — all but five of the petals were stamens. Some of the very narrowest of these imperfect petals actually end in anthers, to let us understand the thing fully. I have already mentioned that stamens, pistils, petals, and sepals are all mere modifications of what were originally foliage leaves, and proof of a very striking character is given in a monstrosity occasionally seen in which all the organs of the flower, petals included, are green. From time to time also botanists record examples of flowers having a leaf or leaves growing out from sepals or petals.

We must now dismiss the Dog Rose, and give our attention to some other forms of Rose-flowers, but to which the names of Apple, Plum, and Cherry are ordinarily attached. In these cases it will be seen that the structure of the flower is essentially the same as in the Rose, with some differences of detail.

Apple Flower

The receptacle, for example, is filled up by the carpels, which are connected with each other and joined to the walls of the receptacle-tube. The five sepals are still attached to the receptacle. The five petals are somewhat different in shape from those of the Rose: at the lower part they narrow into a short claw, and this portion is attached to the calyx-tube just above the base of the stamens. As the five carpels are united, so also are the styles at their lower extremities, but higher up they separate.

Now in the Apple, the Pear, and the Mountain Ash, which are all members of the genus *Pyrus*, the stigmas are mature and ready for fertilisation before any pollen is shed by the anthers, so that it becomes very difficult, though not impossible, to fertilise the seed-eggs by pollen from the same blossom. Another advance upon the Wild Roses is seen in the production of honey by the Apple section of the Rose family. This, of course, is to induce insects to visit the flowers, and the better to enable them to effect cross-fertilisation with pollen brought from another flower or another tree there is this peculiar arrangement of the anthers maturing later than the stigmas.

It will probably occur to some of my readers that under such conditions the Apple-tree that flowers first in any district must be barren, because there would be no pollen available in time to fertilise the seed-eggs. That difficulty is more apparent than real. As a matter of fact all the flowers on a tree do not open at once, not even all the flowers in a small cluster; the central blossom in a cluster opens first, and the pink exteriors of the unopened buds surrounding it serve to make it more conspicuous to insects. Again, if no pollen touches the stigmas, these remain ripe for a comparatively long period, and the petals remain on the flower. It will thus be seen that failing the carriage of pollen by bees from earlier flowers the stigmas may receive pollen from the more backward anthers of their own flower, and it is probable that most of the flowers of our large fruit trees are fertilised by their own pollen, but by insect agency. It is, however, quite clear from this earlier maturation of the stigmas that an occasional cross is desired by the tree.

Except in size and their disposition on the tree—

what botanists term the *inflorescence*—the Pear, the White Beam, the Service, the Rowan, and the Medlar flowers differ but little from those of the Apple; but whereas five of these have their flowers in clusters (*umbels* or *cymes*), the sixth, the Medlar, bears large solitary flowers. The flowers of the whole genus show an advance upon those of the Wild Roses in the fact that they provide honey to attract insects and have adapted themselves for cross-fertilisation. There are other matters in which they also show advance. Growing more slowly and maturing wood, they can stand alone, and do not need any hooked prickles to aid them in climbing; they become small trees. Let us look at their fruits.

Apples

The carpels of the Rose were all separate at the bottom of the tube; those of the Apple were connected by their inner edges, and by their outer edges were attached to the walls of the tube. Now with the impetus given to this part of the Apple-flower by fertilisation the juices flowed freely into the walls of the receptacle-tube until it swelled into a globular form completely embedding the carpels in a thick wall of flesh. The object is the same here as in the growth of the fleshy pulp round the Rose-hip: to attract animals to scatter the seeds. In viewing fruits from this point of view, we must keep our minds clear of the comparatively monstrous Apples and Pears resulting from the artificial selection, grafting, pruning,

Section of Apple-flower

and general attention of the fruit-grower. It is the little wild Crabs (*Pyrus malus*) we have in mind, but, of course, even in cultivated fruits additional size and richness are produced by the same means, guided and controlled by the skill of the gardener. Why do not fruits develop sweetness and soft flesh until they are fully grown? Because that would defeat the object the tree has in producing fruit at all. It is an exhausting process, and it is a common experience for a tree to be all but barren the year after it has borne a good crop. As a tree—a mere matter of wood and leaves—it would probably be a finer specimen if its flowers failed several years in succession. The ripening of fruit does not take place until the contained seeds are fully formed, and then large ones like the Apple are easily detached from the tree, and drop to the ground. Here they are probably seen by some wandering mammal, which is attracted by the colour, and they are eaten —it may be by rabbits, pigs, donkeys, horses, or cows; and then it would appear that the tree's object had been completely frustrated, yet the exact converse is the fact. The carpels (core) are thin, and may be eaten and digested, but the seeds are wrapped up in thick leathery skin which is proof against the action of the digestive fluids for the short time they might remain in the eater's stomach, and by the time they have passed through the creature's intestines they may be several miles away from the parent tree. That is the object the tree had in surrounding the seed-vessels with an attractive flesh. The object of

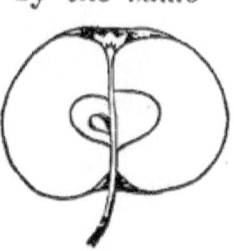

Section of Apple

the Rose was similar, only as the Rose-hip was to be eaten by birds the fruit was arranged to remain standing in a most conspicuous position instead of falling from the bush.

The Wild Pear (*Pyrus communis*) agrees in most details with the Wild Apple, but the form of the first is more conical, the receptacle gradually enlarging from the stalk, instead of suddenly as in the Apple. The smaller fruits of the Wild Service (*Pyrus torminalis*) are variable in shape between that of the Pear and of the Apple, and are small enough to be eaten by birds whole, though their greenish-brown colour would lead one to suppose that small mammals were more concerned in the work of distributing the seeds. A similar remark applies to the Medlar (*Pyrus germanica*), which is coloured brown when ripe, and is of very different shape from the other Apples, the top being flat and the calyx lobes not in the centre. No bird would attack it whilst on the tree, but after it has fallen upon the ground it begins to decay, or "blet," as fruit-growers say, and has then a pleasant acid-sweet taste, which would make it acceptable to field-mice, dormice, voles, etc., who would carry it away and thus scatter the seeds. Another of this section of the Roses is the Mountain Ash or Rowan-tree (*Pyrus aucuparia*), which produces tiny little Apples of a brilliant scarlet in dense drooping clusters

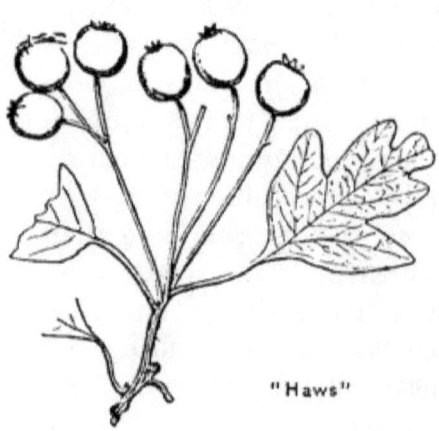

"Haws"

which give the tree a very handsome appearance in autumn, when birds swarm upon it and quickly devour the fruit.

There is yet one other of the Apple section of Roses to which we must refer, and this is perhaps the most familiar and plentiful of all. It is the haw,—the fruit of the Hawthorn or May (*Cratægus oxyacantha*), —and from the outside it is not much unlike the Rowan-berry. Its few-berried clusters stand erect instead of drooping, and its "core," very hard and bony, is composed of one or two, occasionally three carpels, with very little flesh outside.

If we take a glance at the leaves of all these species we shall find great variety in their forms, and this variation appears to be dependent upon the manner of their arrangement on the stem or branch, and the situations in which the trees ordinarily grow; the object being to give the whole area of leaf-surface full opportunity to obtain all the carbonic acid gas possible from the atmosphere, and to get all the available sunshine in order that the chlorophyll, or green colouring matter of the leaf, may act, which it can only do in sunlight. The carbonic acid gas is a compound of carbon and oxygen, and it enters the leaf through the many breathing pores (*stomata*). When it comes in contact with the chlorophyll this seizes hold of the carbon and sets the oxygen free.

The tree is largely built up of carbon, in combination with mineral substances absorbed by the roots, so that it is of vital importance that the plant or tree should get as much leaf-surface as possible exposed to air and sunlight. If you will take a branch from any tree and hold it in just the position it had upon the tree, you will see for yourself how nicely the shapes and sizes of the leaves and the length of their stalks are adapted to secure economy of space without overlapping; and how the thickness of the branch has relation to the strain it has to bear in the weight of the leaves.

Looked at from this point of view, you will find considerable charm in a comparison of the foliage of all these species of Apple-roses. They differ from the Dog Roses in the fact that the leaves are not compound, but simple, some slightly toothed, some strongly toothed, some deeply lobed. One, however—the Mountain Ash—*has* got compound leaves, more so even than those of the Dog Roses, so like an Ash-leaf that it is called Mountain Ash, though having no relationship to *the* Ash. It has been suggested that this departure from the leaf-form of all the other Apples may be due to the windy, mountainous districts it chiefly affects, where a leaf of the normal form and of proportionate size would be torn off by its resistance to high winds, or might lead to the uprooting of the tree or the snapping of its trunk. It is probable that the Rowan has also found small fruits "pay" best, for its little apples are in great demand by birds—so great that bird-catchers use it for the purpose of baiting their snares, and one of its names is Fowler's Service-tree; its scientific name, *aucuparia*, indicates that it is good for "going fowling."

But we must hurry on, for there are many more British Rose-groups that must be considered, simpler in character than these highly-developed Roses and Apples—humble plants which there can be little doubt show us what the original Roses were like, and also indicate the stages of advancement by which the Roses and Apples have come. We have been so far considering the aristocracy of the Rose family; let us now turn to the older but humbler branches of the tribe. On every common and waste place, by every dusty, rural roadside, we may find examples.

This bright, yellow-flowered Cinquefoil (*Potentilla reptans*), which the undiscriminating public sets down as a kind of Buttercup, will serve our purpose as a sample of this group. If a flower of Buttercup be gathered and compared, point by point, with this Cinquefoil, they will be seen to differ very materially, but we may admit that people in too much haste to make such comparisons may easily confuse the two. The Cinquefoil is a little yellow rose, and we can have little hesitation in following Mr. Grant Allen when he claims that the original founder of the great Rose family, from which have ascended the Roses and Apples and Brambles, must have been a plant closely resembling the Cinquefoil. That may appear a bold thing to declare, but there are so many species of plants still in existence which seem to show the way by which the more highly-developed members of the family may have attained their eminence, that it requires less courage to affirm than to deny it.

Cinquefoil

This Cinquefoil has no woody stem, but merely a thick rootstock, from which every year new pink creeping stems of a slender character arise, and these at intervals of a few inches send down roots which bind it to the earth and feed the plant. Above these roots arise the leaves which are compound, but the five leaflets are arranged in a fashion differing from those of the Wild Rose. Those of the Cinquefoil spread out from a common centre, somewhat as the fingers of a hand diverge, and consequently the arrangement is described in the books as *digitate*. Note that though there are fingers to this leafy hand there is no palm. These leaves are on long leaf-stalks, and between the base of the leaf-stalk and the stem (*axil*) the flower-stalks are produced.

Each flower-stalk supports but one flower, consisting of five petals, which are broadest in the middle, whence they narrow greatly to their base. The calyx appears to consist of ten sepals, but five of these are regarded as little bracts (*bracteoles*), forming an *epicalyx*, or calyx upon calyx. Coming to the inside of the flower, we find that the receptacle is not hollowed out like those of the Rose and Apple, but level-topped, both carpels and stamens standing up from its surface. Both ripen at the same time, and as the flowers are much visited by flies, beetles, bees, and butterflies, for honey and pollen, self-fertilisation is freely effected.

Another species of Potentilla is known as the Tormentil (*P. tormentilla*). It differs from the Cinquefoil in having its stems half-erect, not rooting; but there is a variety or sub-species whose stem creeps, and occasionally is known to root. Its leaves are stalkless, and have but three divisions, but the stipules have cut edges like the leaflets, so that the leaf as a whole looks a Cinquefoil. Another peculiarity of the Tormentil which makes it probable that it may be a degraded descendant of the Cinquefoil is the reduction of the petals and sepals to four each, but occasionally it appears with the floral leaves in fives.

Roses and Apples

The well known Silver Weed (*P. anserina*) has flowers similar to, but smaller than those of the Cinquefoil, and its leaves are divided into many leaflets arranged pinnately, or, to be more correct, in an interruptedly pinnate manner, for in between every two large leaflets on each side of the "midrib" there is a tiny little leaflet. The leaves stand erectly, and are silvery on the back.

Yet another Potentilla is called the Barren Strawberry (*P. fragariastrum*). Its leaf is divided into three leaflets only, which are very hairy on the underside and supported on hairy leafstalks. These leaves grow in tufts round the tough branching rootstock, from which also arise flowering stems bearing two or three *white* flowers. Now this plant bears so strong a superficial resemblance to the Wild Strawberry (*Fragaria vesca*) that when they are both beginning to bloom the non-botanical lover of nature fails to distinguish them. But when all these Potentillas we have named have become fertilised by insects crawling over stamens and carpels, and the petals drop away—what do we find? A crowd of little beaked nutlets (*achenes*) on the flat-topped head of the flower-stalk, surrounded by the calyx—all dry and uninviting, and probably trusting to the wind, and the passage of animals over them, to separate them from the receptacle and disperse them, so that the contained seed in each may have a chance to germinate at some distance from the parent plant.

Strawberry-leaf.

Still one more Potentilla we must name ere we pass on. This is the Marsh Cinquefoil (*P. comarum*), which we must seek in boggy ground, where its long purplish stems make a vain effort towards an erect attitude. Its leaves are pinnate, and consist of five or seven leaflets; and its small petals are purplish-brown in colour. This is a striking departure from the usual yellow of the Potentillas, more striking by far than the variation to white of the Barren Strawberry, but it is coupled with another noteworthy difference from its congeners. The receptacle instead of being flat is conical, and after fertilisation this grows enormously until it is as large as a Strawberry, and of a crimson colour, but dry and spongy, without flavour. On this the shiny achenes are borne much after the manner of the Strawberry. I used to grow this as a pot-plant in the greenhouse on account of the ornamental fruit.

The Wild Strawberry (*Fragaria vesca*), we have seen, closely resembles the Barren Strawberry Potentilla, except in the fact that it produces runners which root and establish new plants at intervals. The receptacle also is conical in agreement with the Marsh Cinquefoil, yet with the difference that it develops into a juicy, sweet and fragrant mass in which the achenes are slightly embedded. The flower also exhibits an advance upon those of Potentillas in that it has taken to maturing its stigmas before the pollen is discharged, and so favours cross-fertilisation.

Now let us consider for a moment the probability of the Strawberry originating from a Potentilla—as a matter of fact it is a *Potentilla*, though systematic botanists persist in placing it in a separate genus.

Wild Strawberry

Though some of the Potentillas are tufted and even shrubby, most of the British species show a tendency to prostrate stems, some of them rooting at intervals. Eight of the native species have yellow flowers, two have become white and one purple; one species has a conspicuous enlarged and coloured receptacle. There you have in our paltry dozen of native species (out of one hundred and twenty known to inhabit the world) a sufficient number of tendencies to produce the Strawberry.

Suppose that one like the Barren Strawberry with white flowers and trefoil leaves had accidentally varied in the direction of its receptacle growing large and spongy like the Marsh Cinquefoil, there would be a tendency in its offspring to repeat the—let us say—malformation. If its size and glowing colour attracted the birds that have a weakness for crimson or scarlet, and they ate its flesh with the attached nutlets, the contained seeds would pass through the bird's digestive organs uninjured and be sown in richer soil. This would give the seedlings an advantage, and tend to fix the character by which they had benefited. Then, if the production of a sweet fluid (nectar) on the surface of the receptacle was extended to the interior of the spongy receptacle so that it grew sweet and juicy, this would no doubt cause the plant to be still more sought out by birds "with a sweet tooth," and the seeds to be more effectually distributed. We all know how plentiful the Wild Strawberry is, and how plastic it has proved in the hands of the fruit-grower, who has got from it a very large number of cultivated varieties differing in the size, colour, and peculiar flavours of the fruit.

There are other important groups or genera of this wonderful Rose family, some of which we must consider before we pass to another family. Some of the Potentillas we have been talking of exhibit a tendency to grow erectly, but most of these only succeed in maintaining their slender stems at an angle half-way between the erect and prone positions. One species, however (*P. fruticosa*), is really a shrub. The Herb Bennet or Wood Avens (*Geum urbanum*) has a creeping rootstock from which long stalked leaves spring direct, their blades broken up into three large, toothed leaflets and a number of variable smaller ones. The upright stems are sparingly clothed with smaller leaves, and it is evident that these stems are of a tougher more woody character than those of most Potentillas. The flowers are not very different from the yellow Potentillas, but if cut through vertically the receptacle will be found to be conical, somewhat after the manner of the Marsh Cinquefoils, but covered with carpels ending each in a long, slender style bent at the tip. The flowers produce honey, and the stigmas mature before the pollen is shed, so that there is every chance of cross-fertilisation. Now you may find this plant in abundance on the borders of woods and copses, and along hedgerows. It does not trust to chance for the dispersion of its seeds; it has developed an artful device to get them carried which may be said to be more sagacious than the production of a showy juicy receptacle as in the Strawberry.

Wood Avens

The Avens considers mammals and even man more useful even than birds in this connection, because birds want bribing. After its style has served its proper purpose the Avens hardens it into a delicate bit of flexible wire, turns its bent tip into a more definite hook with polished surface, so that it may be warranted to pass imperceptibly through any textile, or to cling to fur or wool. The receptacle and its nutlets have developed into a relatively large mop-head of bristling hooks, which cannot be touched by coat of sheep, rabbit, dog, or cattle without several of the hooks adhering, and so carrying away the nutlet with its contained seed. Try this for yourself by rubbing such a fruit-head against the clothes of man or woman, and see how readily the hooks take hold. Many of our hedgerow plants employ hooks of one kind or another in the dispersal of their seeds—and that is the reason why they *are* hedgerow plants. These seeds are only to be dislodged by firm pressure; and when the smaller mammals push their way between the stems of the hedgerow, just the right kind of pressure is exerted, and the nutlets drop off. The bright little Agrimony (*Agrimonia eupatoria*) carries its two nutlets in the hardened calyx, whose summit is set around with small hooks which serve a similar purpose.

Agrimony

Now if we turn to the Brambles, which are also Roses, we may get a clue to the probable stages by which they have arisen from more lowly forms in the struggle for existence. The stems of the Potentillas and Geums, being deficient in woody tissue and perishing during winter, are described by botanists as being *herbaceous*; and the two simplest of our native species of Bramble have herbaceous stems, whilst the familiar Blackberry is of a more shrubby nature. If we suppose that the founder of the Bramble branch of the family was a Potentilla or a Geum that took to investing its nutlets with juicy pulp instead of making the receptacle juicy like the Strawberry, I do not think we shall be far wrong. Then in later generations improvements were added on this side and that, until we reach the results seen in our present-day British Brambles. It will be readily seen that if a small plant took to making its nutlets so attractive to birds as we have indicated, it could not so well afford to produce so many; indeed, the greater certainty of distribution would render wholesale production quite unnecessary. Our simplest species is the Stone Bramble (*Rubus saxatilis*), which grows on the rough stony banks of mountain streams and copses in the western half of Britain, and in Ireland. It has a creeping rootstock which sends out runners like those of the Strawberry, and from these shoot up erect flowering stems a foot or so in height. Its leaflets are in threes, covered with soft down, and the stems are protected by a few bristles, which are more numerous on the flowering shoots. The few flowers are very poor affairs for Roses, the greenish-white petals being small, narrow, and but slightly opening, consequently it is not well patronised by insects. The stigmas mature before the anthers. It is mostly fertilised by bees, but frequently also by its own pollen, the numerous stamens bending over the two or three styles to deposit their pollen. The carpels at the base of these styles contain two seed-eggs each instead of one, as in Potentillas, but as a rule only one develops; and as the carpel grows, its juicy covering grows as well, so that we have not a nutlet or *achene* in the fruit, but a *drupe*. The final product of the Stone Bramble's flower, then, is a solitary juicy scarlet drupe, or a cluster of two or three drupes.

The other herbaceous Bramble is the Cloudberry (*Rubus chamæmorus*), which grows on high peaty moors in the north of these islands. It is of somewhat similar habit to the Stone Bramble, but its creeping, branched rootstock does not give off runners, and its stem, though growing erect, is only about half a foot in height. Nevertheless, its solitary flower is an inch across, with white petals, and its compound fruit consists of nearly a dozen round juicy drupes, at first of a bright scarlet colour, changing when ripe to rich orange.

Cloudberry

In these two species the stems are covered only with down and bristles, which serve to prevent ants from climbing to the flowers and taking away pollen or nectar without rendering service in return. The Wild Raspberry (*Rubus idæus*) has tall, erect-growing, tapering, shrubby stems, which last for two seasons, and these are densely armed by slender prickles, which readily pierce the hand, and of course prevent the stems and leaves being devoured by cattle. The flowers are in small drooping clusters, but the white petals are small and narrow, so that the blossoms are not particularly conspicuous, and are but little visited by insects. Honey is produced by a fleshy ring near the base of the stamens, but is not very accessible. Both anthers and stigmas mature together, and self-fertilisation frequently takes place. As the numerous and closely-packed drupes ripen and become yellow or red, they separate from the long and slender spongy receptacle, so that when the fruit is gathered it is free from the calyx, thimble-shaped, and hollow. The habit of this species should be noted: the erect stems arise in a group from the rootstocks, new shoots, or "suckers" as they are called, continually arising round the old ones, which die off in turn.

Blackberry

The Bramble, or Blackberry (*Rubus fruticosus*), does not produce suckers, and its shrubby stems are more woody and persistent, covered with stiff, sharp bristles, gland-tipped hairs, and strong curved spines. With the aid of these hooks it has become a climbing plant, clambering between the close branches of other bushes and hedgerow trees, even making its way up the rough bark of a tree and rooting among the vegetable mould that accumulates where the trunk forks. It is quite a common thing for a Bramble stem to grow so long that its own weight bends it over and it forms an arc; the growing tip roots in the ground, and then sends up new stems. In this way it rapidly occupies much ground. The white or pink flowers are borne in clusters of varying length, and their large petals and numerous stamens render them conspicuous; they bear honey, and are much visited by bees and butterflies, as well as by flies and small pollen-stealing beetles. Hermann Müller has enumerated close upon a hundred insects of various orders that visit the Bramble blossoms for honey or pollen. The stigmas and outer anthers mature together, but the anthers hold themselves as remotely from the stigmas as possible, so that the pollen becomes easily accessible, and there is every chance of a pollen-dusted bee from another bush settling on the central stigmas and fertilising them before any pollen from the same flower has had time to reach them. If cross-fertilisation has not already taken place, self-fertilisation will be effected when the inner stamens mature and raise their anthers to a level with the stigmas. As everybody knows, the fruit of the Bramble when ripe is black or very dark purple.

The Bramble cannot be said to have reached the limits of development, for it is one of the most variable plants we have. Even in our little British Islands there is so enormous a number of variations of the Bramble, that botanists almost despair of them. By some authorities they are reckoned as sub-species of *Rubus fruticosus*, each with a number of varieties, whilst others regard most or all as species distinct from *R. fruticosus*. I consider the connecting links are too close and too numerous to so regard them, but no doubt in the course of time many of these connecting varieties will disappear, and the future botanist will then have little difficulty in separating and identifying those that are left. It is in some such way that our existing species of plants have been created, and this multiplicity of Bramble-forms helps us to understand the process.

If we were to imagine the Stone Bramble varying in the direction of putting all its juiciness into a single drupe (as it frequently does), and to largely increase the size of both drupe and carpel, making the latter thick and bony, we should have in effect a—*cherry!* A cherry is a large drupe containing a single carpel (the cherry-stone), and suspended by a long stalk; a plum differs little save in its more elongated shape, its flatter stone and shorter stalk. The flowers of plums and cherries are also roses!

Sloe

We have four native species of these "stone-fruits," and of these the first is the Blackthorn (*Prunus communis*), producing the Sloe as its fruit; and the sub-species *P. insititia*, producing the Bullace, and *P. domestica*, whose fruit is the Wild Plum. The Blackthorn is a much-branched shrub, with hard tough wood, each of the branches ending in a sharp spine, caused by the withering of the growing point. In common with the Whitethorn or Hawthorn, previously referred to, the Blackthorn has learned this trick of protecting itself from being eaten up by cattle,—a stab by one of these thorns on a sensitive muzzle being sufficient to discourage horse or ox. The white blossoms are small, but, appearing before the leaves have unrolled, are very conspicuous in contrast with the black branches. The calyx has five lobes, and the petals are five; the stamens are numerous, but there is only one carpel, terminating in a long style, and containing two seed-eggs, of which as a rule only one grows to be a proper seed. The flowers are honeyed, and the stigma is ripe for fertilisation before the pollen of that flower is shed,—in fact, before the flower is fully open,—but should no visiting insect bring pollen from farther along the hedgerow, crossing may be dispensed with. The smaller bees (*Andrena*) and flies are the fertilising agents. As a result of fertilisation, the walls of the ovary swell and thicken with pulp as the carpel enlarges, and both calyx and corolla drop off; and five or six months later it has developed into a round black plum, thinly covered with a mealy wax ("bloom"), which gives it a bluish appearance. It is probable that this wax is secreted for the protection of the fruit until it is ripe, either in keeping off dampness or making the skin objectionable to insects that would otherwise attack it. From the Sloe through its sub-species our cultivated plums have been evolved by artificial selection.

Cherry Blossom

Our Dwarf Cherry (*P. cerasus*) is something

between a bush and a small tree, partaking of the nature of both. It has a red bark, drooping branches, and erect, coarsely-toothed dark-green leaves, which appear with the flowers or just before them. These flowers are very like those of the Blackthorn, but larger, and the notched petals are less widely spread. They stand erect on long stalks, either singly or in bunches (*umbels*) of two to four. The little red cherries are round, and their juice very acid.

The Gean (*P. avium*) is a real tree, forming a stout trunk, twenty, thirty, or even forty feet in height. Its branches all take an upward direction, and its large pale-green leaves hang downwards. Its soft petals fall widely open. The fruit may be red or black, and is not so round as the Dwarf Cherry, the part adjoining the stalk being broader than the other end. In both these species both stamens and pistils mature simultaneously, but in the third species, the Bird Cherry (*P. padus*), the pistils are earlier. The flowers are grouped in *racemes*,—that is, the flower-stalks arise at intervals from a longer central stalk; the petals have somewhat ragged edges. At first these flowers are erect, then droop. The fruit is egg-shaped, black, and erect. Its flavour is very bitter. From the first two of these cherries all our marketable kinds are believed to have been cultivated.

Cherries

Now, these succulent fruits have all been evolved by the Brambles, Plums, and Cherries to please the birds, in order that thereby the plants' end might be served. That end once more is the dispersal of the seeds. If plants were animals, it would not be considered ridiculous to imagine the progenitor of these single-seeded shrubs and trees arguing that now the plant had attained such large proportions it was not necessary to cover itself with blossoms, each producing twenty or more seeds. "Let us produce a goodly number of seeds, but make them larger, so that on germinating they may more rapidly grow big and strong; one or two seeds to a flower, and the value of the material saved expended in a good thick attractive coat that will tempt birds to swallow them whole, without thinking to look for the seeds." Then, of course, the seed had to be protected against digestive fluids, for the fleshy part of the fruit being soft and juicy would soon be dissolved. And so the carpel was made thick and hard, strong enough to resist digestion and pass out with the seed uninjured. To taste first these wild Cherries and Plums, and then the Bramble fruit, will give us a notion of the varying tastes of birds. That some of them should enjoy acid, bitter, and austere fruits of this kind, should not, however, occasion any surprise, for we know that some poisonous (to mammals) berries, like those of the Arum and Belladonna, are eaten by certain birds without ill effects.

Several minor genera of British Rose-worts have not been mentioned, though they have an interest of their own as examples of probable degeneration from the typical forms. Among these are the beautiful Spiræas, the Lady's Mantles (*Alchemilla*), and the Burnets (*Poterium*). Our two native Spiræas are the well-known Meadow-sweet (*Spiræa ulmaria*) and the Dropwort (*S. filipendula*). The first-named growing in wet meadows and along the margins of streams, where its beautiful much-divided leaves and its rich creamy plumes of fragrant flowers form so striking a feature of meadow vegetation. Individually the flowers are small, therefore the plant has had to follow the example of the Hawthorn in producing enormous numbers of them; and here they are clustered closely in what are termed compound cymes. The flowers are not honeyed, but they are very fragrant, and they produce pollen on a scale of great extravagance, if we merely keep in mind the chief use of pollen. There are only about half a dozen carpels in a flower, and a couple of anthers would supply pollen enough for these; but then insects would soon find out the fraudulent pretence of perfuming the air yet offering no refreshment to visitors. So the plant has kept up a large number of stamens which produce abundance of pollen, and beetles, flies, and pollen-seeking bees are rewarded for carrying a little of it from plant to plant. As the anthers shed their pollen before the stigmas are mature, there is every chance for cross-fertilisation. The Dropwort is of more lowly growth, and affects dry pastures and downs. Owing to this difference of habitat and the liability to suffer in dry seasons, it is not surprising to find that the Dropwort has adopted the precautionary measure of laying up moisture in little tubers it has formed on its root-fibres. In this plant both anthers and stigmas mature together.

Dropwort

Dropwort Flower

The three native species of Alchemilla have tiny

Meadow Sweet.

inconspicuous flowers, the corolla being entirely absent; but it is very probable that they have seen better days, for they still produce honey. Most of the flowers are deficient either in stamens or pistil, one or the other being aborted, and this condition of course brings about cross-fertilisation through the visits of short-lipped insects.

The two species of *Poterium* also lack petals, and but for the fact that the flowers are clustered together in rounded heads at the top of the tall flower-stalk, they would be unnoticeable. As it is, they are liable to be passed by most people as plantain flowers. Closely examined, there will be found considerable difference in the flowers of the two species. The Salad Burnet (*Poterium sanguisorba*) in most cases keeps its stamens and pistils apart; the upper flowers in a head producing a pistil only, the lower ones stamens only, or occasionally with a pistil among them. The plant has, so to speak, turned its back upon the insects, and laid itself out for cross-fertilisation by wind-agency,—hence its long style branches at the summit into a perfect brush of stigmas to catch the flying pollen-grain; hence, also, its twenty or thirty excessively long stamens, that its extravagant output of pollen may be caught by the wind and swept away. There are only four calyx-lobes instead of the usual five, but in spite of this fact there is so close a similarity in the top-shaped receptacle-tube of *Poterium* and *Agrimonia*, that it is reasonable to suppose that *Poterium's* distant ancestor was an ancestor of *Agrimonia's* also, and that it had yellow petals and was insect-fertilised. For this purpose a dozen stamens or less sufficed; but the Salad Burnet, afflicted with a craze for cheap labour, wooed the wind, and soon had to give up its gay golden petals in order to provide longer stamens and pistils and more abundant pollen. There can be no question of the desire of this plant for cross-fertilisation, because from the way it has kept its pollen-bearing flowers below the seed-bearing ones any other method is scarcely possible. The Great Burnet (*P. officinale*) appears to have made a strong effort to get back to friendly relations with the insects,—perhaps from having found that the stations it had taken up in damp meadows were not so favourable for wind-fertilisation as the high-lying downs and pastures frequented by its smaller relation. Its petals, alas! were gone, but it economised in the pollen department, —cutting down the number of stamens to four, and making these so short that they do not stand above the calyx-lobes,—simplified its stigmas, and enlarged its calyx, putting some purple colour into these at the same time, and—more important still—took to offering nectar to its insect-friends.

BUTTERCUPS AND COLUMBINES

QUITE early in the new year the botanist who is looking along the hedge-banks and the copse-sides to see how things are moving, is sure to come upon great numbers of neat little glossy, heart-shaped leaves, not much more than half an inch long, some of them with whitish patches upon them. They lie pretty flat upon the ground, or a little above it, their stalks radiating from a central rootstock. These are the leaves of the Lesser Celandine, or Pilewort (*Ranunculus ficaria*), and a week or a month later we may find, in addition to the leaves, an abundance of starry blossoms of rich burnished gold flashing brightly in the fickle sunshine. If, when the flowers have made their appearance, we take up one of the plants with a trowel and wash the earth from its roots, it will serve as an introduction to the important Buttercup family (*Ranunculaceæ*).

The roots of the plant are seen to be thick, fleshy,

and suggestive of Dahlia roots. This is due to the storing up in them of the food supplies manufactured in the leaves of the plant last year. When the heat of summer came, drying up the moisture the Pilewort had found so useful in the spring, there was danger that so soft-textured a plant would be dried up and destroyed. As a fact, its stems and leaves withered and disappeared, but that was due to the cuteness of the plant in transferring all its substance to its root-fibres, which expanded their cells to accommodate it. All plants that are able to indulge in very early and sudden displays of leaf and flower do so by the practice of thrift in the previous year. They do not throw off leaves charged with plant-food, as do many of the trees and shrubs; the leaves instead are gradually emptied of all that is worth saving, and so the exposed portions of the plant *wither*. This is the process adopted by all the bulbous plants,—Crocus, Tulip, Hyacinth, etc.,—and by drawing upon their savings such plants can blossom as soon as the severities of the winter have passed, and before their leaves have been able to accomplish their proper functions. Ages ago, when men held what was called the Doctrine of Signatures, this bunch of thickened root-fibres depending from the base of the Celandine was thought to resemble hæmorrhoids, therefore they gave the plant the name of Pilewort, and considered

Petal, with Nectary

Lesser Celandine

the likeness to be an indication that it was a cure for the complaint. Hear old Nicholas Culpepper on this point:—

"Behold here another verification of the learning of the ancients, viz., that the virtue of an herb may be known by its signature, as plainly appears in this; for if you dig up the root of it, you shall perceive the perfect image of the disease which they commonly call the piles. It is certain, by good experience, that the decoction of the leaves and roots doth wonderfully help piles and hæmorrhoids, also kernels by the ears and throat, called the king's evil, or any other hard wens or tumours. . . . . The very herb borne about one's body next the skin helps in such diseases, though it never touch the place grieved."

Lesser Celandine's fleshy roots

Before the first flowers appear, the Celandine lengthens its stem slightly, and the leaves that grow from the stem vary from those that grow from its base, in the fact that they are more or less acutely lobed, and thus approaching somewhat to the shape of ivy-leaves. They have rather long leaf-stalks, which are widened at their base so that they can clasp the stem. The stem branches from the lower part, and at intervals bears solitary flowers on long stalks. These flowers, though not greatly unlike certain Potentillas at a superficial glance, will be found to differ considerably on closer examination. We saw in the Rose family a great tendency to vary in the direction of size, consistency of stem, form of leaf, and so forth, but very few departures from the regular five-parted calyx and corolla. The Ranunculus family is also conformable in the main to this *pentamerous* arrangement, as it is called, but we shall find all sorts of departures from it, so that one commencing the study of botany may well be puzzled for a time by the apparent chaos of arrangement. Celandine would be regarded by most botanists as a bad example to select as a type of the family, but it is a good type at least in the sense that it exhibits the erratic tendencies of the group. It should have five sepals, but usually there are only three, sometimes four or five; it should have five or ten (5 × 2) petals, but usually they are seven, and may be any number up to twelve, or, as sometimes happens, they may be entirely wanting.

Section of Celandine

The petals narrow to their base, where there is a honey-gland. There are many stamens (A) of varying number, and the smooth carpels form a globular head. Most of the stamens shed their pollen before the stigmas (ST) are ripe, so that the bulk of the pollen is intended for the fertilisation of other flowers with the aid of the numerous flies, beetles, bees, and other insects that come after honey or pollen. As the anthers successively mature, they turn as far away from the stigmas as possible, so that they have every chance of being dusted by pollen shaken from the bodies of visitors who have been to other flowers previously. By the wandering of the smaller insects from anthers to stigmas, there is no doubt that fertilisation is frequently effected. The carpels develop into achenes not much unlike those of the Potentillas; but it has other means of reproduction, for frequently little tubers form at the base of the leaf-stalks, and as the plant fades these strew the ground in places, and each produces a new plant next year. As these tubers are about the size of grains of wheat, their presence on the ground has sometimes led people to imagine that it has rained wheat! The swellings on the roots are also detachable, and each capable of producing a new plant.

It is interesting to consider the probable cause of the tuberous roots in Celandine. There can be little doubt that it is one of the oldest forms, if not *the* oldest form of Ranunculus we have—the shape of the lower leaves testifies to that. Perhaps Celandine and Serpent's-tongue (*Ranunculus ophioglossifolius*)— now all but extinct in this country—exhibit the branching off from a common ancestral form, the former towards Marsh Marigold and Globe-flower, the latter to the Buttercups through the Spearworts. Celandine was then probably an annual, as Serpent's-tongue is still, and flowered in summer, as most of the family do. We can imagine Celandine growing as now under the shelter of bushes, but not being so pressed by competitors for space and light. Some changes brought about the crowding of such places by hogweed, fool's-parsley, and other broad-leaved plants, which threatened to suffocate the little Celandine, and prevent it getting the materials necessary for flower production. Three courses were open to it if it was not to be snuffed out,—it must change its habitat, grow a tall tough stalk, as the Meadow Buttercup has done, that its flowers might flourish above the crowd, or it must change its flowering season.

The annual Celandine adopted the last course, and in so doing became a perennial. Its seedlings were no longer to labour to flower during their first season; they were to work and save all they could, packing it into those root-swellings. Then the second year the young plant started with a full exchequer, and could afford to spend its substance in a floral display as soon as winter had passed, and before its competitors for place had time to develop their leaves. The earliest insects fertilised its seed-eggs, which were fed by the ample stores below, and the plant had still time to enlarge its leaves and let them make abundant food before the summer fully set in with its overcrowding. Then it scattered its seeds, drained its leaves and stems of all their wealth, and packed it into the underground treasury in readiness for the new year's expenditure.

There are several species of Ranunculus that are popularly lumped together under the name of Buttercup. They are of taller growth than the Celandine, have larger leaves, deeply lobed and cut at the edges, the whole plant more or less hairy, and the petals broader, so that the flower forms a golden cup instead of a star. The Celandine grows chiefly under the shadow of trees and bushes where there is no other vegetation, at least during its flowering-time, so that there is little occasion for a tall stem. The Buttercups grow chiefly among grass and weeds, where the competition for light, air, and the attention of insects is greater; therefore, as the surrounding vegetation grows, they have to grow taller. Ants are notorious honey-robbers if they have the chance of getting at honeyed flowers, but they are still enjoying their winter sleep when the Celandine flowers, so that there is no need for the plant to adopt defensive measures against them. The Buttercups, on the other hand, flower later, and to keep their flowers safe from insects crawling up from the ground, they have covered their stems and leaves with hairs, because they invite the attentions only of those insects that are able to fly from flower to flower and carry pollen on their hairy bodies. Even if ants could fly, but little pollen could attach to their polished bodies.

It appears from a general survey of the plant world, that these lobed and slashed-edged leaves are more useful for plants that grow a little distance from the ground, whilst those that grow close to it in more open spaces can get on quite as well with leaves of simpler form. It is quite clear that any plants growing in meadows and pastures, or along the grassy roadsides, are very liable to be eaten up by herbivorous animals. The grasses have adapted themselves in various ways to endure this browsing without being extinguished, but their story we must leave till their turn comes; the Buttercups have solved the difficulty by developing an acrid, poisonous principle in their leaves and stems, so that their juices will blister the mouths of cattle. It used to be said that healthy tramps, in order to touch the susceptible hearts of the charitable, rubbed their naked feet with *Ranunculus acris*, whereby they raised blisters which were palmed off as evidence that they had walked many miles in their everlasting fruitless search for the work they hope never to find. Little children, too, have had their delicate hands blistered by tightly grabbing the bleeding stalks of Buttercups they have gathered with such joy. Let your eye range over a pasture at midsummer, when the grass is very closely cropped by sheep or cows, and see how strongly the dark-green clumps of Ranunculus stand up above the turf. They have been avoided by the animals. Celandine does not develop this acridity, neither does the Water Crowfoot, which grows out of the way of browsing animals.

There is one point in the structure of these Buttercups to which attention should be called, because botanists rightly attach considerable importance to it. If a flower be cut vertically through the middle, it will be found that the receptacle is a lengthened cone from which sepals, petals, stamens, and carpels successively spread out. It will be clearly seen here that the stamens are attached directly to the receptacle and not to the sepals. Space will not allow us to deal with the many native species of Ranunculus, but we can briefly glance at the differences between a few, to indicate to the reader the kind of variation to look for in the others. The Upright Buttercup (*Ranunculus acris*) has an erect stem, much divided leaves, sepals and petals spreading widely, the anthers shedding their pollen before the stigmas are mature, and, though capable of self-fertilisation, largely crossed by many kinds of insects, chiefly beetles and flies of brilliant colours, who are attracted by the shining golden hue of the flowers.

The Bulbous Buttercup (*R. bulbosus*) is also erect-stemmed, but not so tall, the base or rootstock swollen into a corm like that of the Crocus but larger; the leaves are less jagged, more distinctly divided into three parts, the flower-stalk is grooved, the sepals bend right back and press against the stalk, whilst the petals remain erect so that the flower assumes a true cup-shape. The Creeping Buttercup (*R. repens*) is somewhat erect, but its lower portion *leans* on the ground and sends off runners, the leaves are divided into three wedge-shaped and toothed segments, the flower-stalk is grooved, the sepals spread widely, and the petals are half-erect.

Bulbous Buttercup

Then there are the Greater and Lesser Spearworts (*R. lingua* and *R. flammula*), which grow in marshy places, and whose leaves are undivided, long, and narrow,—a form common in marshes, ditches, and the borders of ponds; the woodland Goldielocks (*R. auricomus*), with roundish three-lobed leaves, lacking the acridity of its relations, and with yellowish sepals to the somewhat irregular flowers; the Water Crowfoot (*R. aquatilis*), with submerged leaves reduced to mere thread-like dimensions, whilst the floating leaves are divided into three broad lobes or leaflets, and the petals of the flowers are white with the exception of a little yellow near the base, which is an indication to flies and bees and beetles that there the honey lies. The thread-like leaves are common to several aquatic plants having no relation one to the other, and show how similar surroundings cause like adaptations in plants differing widely in structure. Such minute division enables the plant to expose a large surface to the carbonic acid gas dissolved in the water, without a large expenditure of material.

Some insects alight on the petals of the Buttercups, then crawl over the anthers and stigmas, from which they reach down to the honey-glands, and so often effect self-fertilisation; others come straight to the centre and bring pollen from an earlier visit, and so cross-fertilisation is assured.

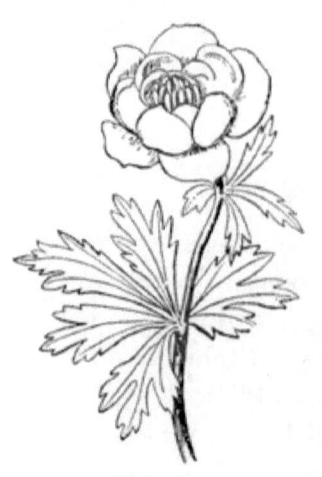

Globe-flower

In pastures and woods at a considerable elevation there grows a glorified Buttercup with leaves divided like the fingers and palm of a hand, and with large globular pale-yellow flowers. It is the Globe-flower (*Trollius europæus*), and its solid-looking blossoms owe their fine appearance to the concave sepals, which may be any number from five to fifteen, enlarged and coloured to look like petals (*petaloid*). The petals are there in equal number, but reduced to narrow, oblong bodies that might pass muster among the numerous stamens, which are of similar length.

They are, of course, easily distinguished by the absence of anthers at one end and the presence of the honey-gland at the other. The carpels are not unlike those of the Buttercups, but here, instead of developing into one-seeded achenes as there, they become *follicles*—large pouches, open along the upper side and containing many black seeds. Stigmas and anthers mature together, and fertilisation is effected by flies, beetles, saw-flies, and bees.

Petal of Globe-flower

Of somewhat similar character is the well-known Marsh Marigold (*Caltha palustris*), that grows in marshy ground and on the banks of backwaters. With its large glossy, heart-shaped leaves and open golden flowers, it is perhaps more suggestive of an enlarged Celandine, but as in Globe-flower the showiness is produced by the sepals, for here there are no petals at all, so the honey is produced at the sides of the carpels. It is a coarse-growing plant, and after fertilisation the leaves develop to an enormous size, but they are kept small during the flowering period, so that the blossoms may not be hidden from the flies, beetles, and bees that are wanted to fertilise them. It is to the enlarged sepals, again, that the two native species of Hellebore (*H. fœtidus* and *H. viridis*) owe such small amount of publicity as they can command. It is true they possess petals, and a goodly number, too; but these are singularly turned into two-lipped tubes containing honey. The stigmas mature before the anthers, so that cross-fertilisation is probably effected—I cannot say by what insects, but most likely by early flies, judging by the colour and odour of the blossoms.

In the Wood Anemone (*Anemone nemorosa*) we have flowers of considerable beauty, entirely due to the six or more sepals, which have been enlarged and coloured white, sometimes tinted with pink, purple, or blue, whilst the petals are entirely unrepresented. The rootstock is cylindrical and tough, creeping just below the mossy soil of the copse and woodland, sending up separate leaf-stalks and flower-stalks. The leaves are divided into three wedge-shaped, cut-edged leaflets, and the flower-stem bears three similar leafy bracts half-way up to the solitary nodding flower. There are many stamens, and numerous carpels, which become one-seeded, downy achenes. Both sets of organs mature simultaneously, and although no honey is offered, bees visit the flowers, attracted by their colour, pierce tender parts of the flower, and suck up the exuding juice. Cross- and self-fertilisation occur indifferently.

Wood Anemone

It would appear to be the aim of the Buttercups and some of their relations to have the bright colours of their flowers seen from below as well as from above, for there is a tendency to throw off their sepals soon after expansion, so that they may not obstruct the view of the shining gold. These yellow flowers are especially attractive to small beetles, which fly low and range over the foliage, and it is probably of service to the flower that these should be able clearly to discern it from below. Therefore, a plant that accidentally developed the yellow colour in its sepals would get an advantage owing to its greater distinctness from below, and some existing species — Goldielocks, for example — exhibit this tendency to yellow sepals. On the other hand, Celandine, which grows low, wishes to be seen from above; so the under side of its petals is almost entirely green, developing into purple, and throwing up the gold of its upper surface with greater strength. This darker colour has also the advantage of preventing the heat-rays from passing through the sepals and away,—an important matter to plants that flower so early in the year.

In other flowers we have seen that the sepals have become entirely coloured and enlarged, so that they are exactly like petals, and are so regarded by most persons. With this promotion of the sepals, the petals lose their importance, and they have been reduced in size, converted into nectaries, or dispensed with altogether, as in the case of the Marsh Marigold and the Anemone. When the Anemone-bud first appears, it is sheltered by the leafy bracts (*involucre*); but when the bud expands, the upper part of the slender flower-stalk lengthens so that the flower is carried high above the involucre, and bends to the breeze. Although the flower has no honey, it is delicately scented, and no doubt when this perfume is borne upon the breeze, and insects are coming up in response, it is an advantage that the face of the flower should be turned in that direction.

There is one other native species of Anemone, and that is the Pasque-flower (*Anemone pulsatilla*), so called because in other days country people used to employ its juices in staining Easter eggs (Pasque eggs). But it is more interesting to us because its flowers show some advance upon those of the Wood Anemone. Its leaves, which are cut up into many very slender divisions, do not properly develop until after flowering. The flower-stalk is stouter than that of the Wood Anemone, and clothed with fine silky hairs. Its six sepals are also silky, and of a dull purple tint. The carpels, instead of ending in a short simple style as do those of the Anemone, have long feathered styles, and in the ripe achenes these are an inch and a half long, and help in the dispersion of the seeds. Now, there is no doubt that the deeper colour of the sepals has made the Pasque-flower attractive to a larger number of insects than patronise the Anemone, for the former has found it possible to do without some of its stamens as pollen-producers, and has actually converted the outermost row into nectaries for the greater encouragement of its winged patrons! The honeyless Anemone, to make sure of getting fertilised, ripens its anthers and stigmas together, but the Pasque-flower with its honey-bait is more certain of visitors, and can arrange for cross-fertilisation by discharging its pollen before its stigmas are mature.

There is a small group of three native plants called

Pasque-flower Stamen and Nectary

Wood Anemone.

Meadow-rues (*Thalictrum*), scarcely known except to botanists, though their pretty foliage entitles them to consideration. They are plants which — though members of a bright-flowered family—teach man that his favourite theory that the chief end of flowers is to please his eye is all moonshine; for these have been so forsaken by insects that the flower has had to press the wind into its service as a pollen-carrier, with all the waste that involves. Flowers fertilised by wind-agency (*anemophilous*), with bright-coloured congeners, may be generally regarded as degenerate, and this is probably the course of their downward career: they made a coloured calyx do instead of a corolla, which completely vanished, and with the petals probably went the honey-glands; then the insects gave up their visits as unprofitable to themselves, and only those flowers that chanced to fertilise themselves contrived to set seeds. Lack-

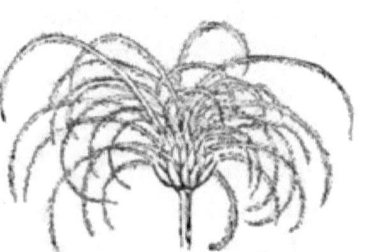

Seed-head of Clematis

ing the stimulus afforded by insect irritation, less nutriment flowed to the sepals, and these dwindled and began to revert to the purplish-green or yellowish-green that is a sign of floral poverty. The essential organs would increase in length, the stamens producing much pollen and the stigmas maturing before the anthers. And so we find them to-day, the yellow anthers the most striking feature of the flowers. The Yellow Meadow-rue (*Thalictrum flavum*), though honeyless, attracts pollen-feeding flies and honey-bees by the attractiveness of its yellow stamens; but the Small Meadow-rue (*T. minus*) appears to be entirely dependent on the wind.

Our solitary native Clematis, or Old Man's Beard (*Clematis vitalba*), is very interesting, from the fact that alone among British members of the family it has developed woody stems and become a climbing shrub. It has no proper tendrils like the Vine, nor hooks like the Bramble, with which to climb, but it has hit upon quite as good a means of rising in the world. Its leaves are broken up into about five distinct leaflets, each with a foot-stalk, and when these touch against a branch of the hedge-trees or other likely support, the foot-stalk intelligently takes a turn round it, and becomes hard and woody. The flowers have lost their petals, and the four thick downy sepals are coloured greenish-white. They produce no honey, but give out a perfume which attracts flies. The numerous stamens mature before the stigmas, and erect themselves so that they occupy the centre of the flower when the anthers discharge the pollen; but the pollen all shed, the stamens droop, whilst the feathered stigmas lengthen and occupy the former position of the stamens, so that an insect coming from a younger flower with pollen on its under side and legs will alight on the stigmas and fertilise them. From the elevated position of the flowers and the feathered character of the stigmas, it is probable that the plant is partially anemophilous.

The inconspicuous little Mousetail (*Myosurus minimus*), found in cornfields, is worthy of a brief note, on account of the remarkable manner in which it fertilises a large number of carpels with only about five anthers. The entire plant is only a few inches high, and consequently easily overlooked. All its leaves are radical,—that is, they arise directly from the rootstock. They are narrow, spoon-shaped, and somewhat fleshy. The minute flower is solitary at the summit of a tall stalk, with five narrow sepals and an equal number of tubular greenish petals. The carpels are very numerous, attached round a slender spike. When the flower opens, this spike is scarcely longer than the stamens, but it begins to lengthen, and continues to do so, rubbing the stigmas in succession against the anthers, and so fertilising them all with a minimum number of pollen-grains. When the process is complete, there is a long spike of densely packed achenes, three inches in length, which has given the flower its name. We shall find no quite similar instance among our flora. There can be little doubt that this is another instance of degeneration from an insect-fertilised condition,—indeed, some of the earliest carpels may still be fertilised by the few flies that visit the flower, but the maintenance of its petals, though sadly reduced, and the small basal spur to the calyx, point to the probability that it once secreted honey in that spur and so made it worth while for insects to visit the flower. The lengthening of the floral axis is probably only one of the shifts to which poverty subjects plants as well as men.

Let us pass on from these poor relations of the Buttercups to the well-to-do members of the family,— those that hold their heads high, having attained to the dignity of blue finery, and intimated that their transactions with the insect world must be restricted to the intelligent and prosperous bees. There is no necessity to describe the Columbine (*Aquilegia vulgaris*) beyond the flower-parts, for its general cultivation in gardens renders its form and foliage familiar to many who have never met with the wild plant. The flower is well equipped for insect-fertilisation, for its five sepals are petal-like and coloured, the five petals are concave and each developed backwards into a hooked spur in which honey is produced, and the numerous stamens mature before the five stigmas. The flower droops, and when it opens the bunch of stamens are found lying to the lower side of it, with the exception of a few that have elevated themselves so that they occupy the centre of the flower. As they mature in succession the remaining stamens attain this position, and finally when the stamens have all shrivelled, the stigmas are left in possession of the centre. Long-tongued bees are the fertilising agents, and they regard the stamens as a convenient alighting-stage from which they can push their tongues into the petal spurs and extract the honey from the hollow knob at the very end. And when they retire their under side is covered with pollen, which is rubbed against the stigmas when they visit an older flower. Cross-fertilisation *must* take place in this case. When one considers the angle at which these flowers hang, the hollow knob at the end of the spur is seen to be a necessity; without it, the honey would drain out to the mouth of the tube, where any vulgar little fly could lap it up without earning it. The Common Humble-bee (*Bombus terrestris*) cannot reach the honey legitimately, but has learned to bite holes in the spur, near the nectary, and suck it without earning it.

There is a somewhat similar arrangement in the Larkspur (*Delphinium ajacis*), but the blue, pink, or white flowers have undergone some modifications, becoming irregular and less perfect. There are still five coloured sepals, one of which has developed a long tubular spur, but there are only two petals, each with a spur which is laid inside that of the sepal, where it secretes honey. At their lower part these petals come close together, but leaving an opening to the spur higher up. The lower sepals constitute an alighting-place for bees whence they can climb the lower part of the petals and push their tongues down the tube to reach the honey, but as in Columbine the stamens and finally the solitary stigma are elevated in succession, so that they come in the way of the bee's proboscis, and cross-fertilisation is thus effected. In some other species of Larkspur the employment of four petals more effectually bars the way to the spur from below.

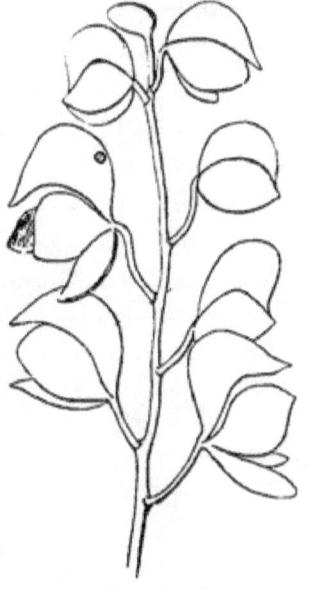

Monkshood

Monkshood (*Aconitum napellus*) is specialised to the same end as Columbine, yet with an entire difference of form. The five sepals are coloured deep blue and are of varying shape; one which takes the uppermost position in the very irregular flower has become an arched hood large enough to contain all the others in the bud. There are only two petals, and these have been converted into hammer-shaped nectaries hidden in the monk's hood. The whole flower bears considerable resemblance to the closed helmet of the knights of old. The entrance is partially blocked by the depressed stamens, which offer a clinging-place to the Garden Humble-bee (*Bombus hortorum*), by which alone the flower appears to be visited. The anthers rise successively, shed their pollen where the under side of the bee will pick it up, then curl back out of the way. The stigmas then elongate and come in the way of the bee's abdomen.

Section of Monkshood

These three higher types of the Ranunculus family have lost the power of fertilising themselves, and as they have laid themselves out so entirely for the patronage of the larger bees, they run great risks of becoming extinct. Their carpels develop into open follicles containing many seeds, but they are yet not common flowers in this country,— certainly not nearly so plentiful as the simpler flowers that fertilise themselves, or as those which, while laying themselves out for insect-fertilisation, reserve the right, so to speak, to fertilise themselves if the appropriate insects do not call in time.

Anthers of Monkshood
Those curled back have shed their pollen

Anthers all curled back out of way of ripe stigmas

No doubt in some seasons adverse circumstances keep down the number of humble-bees, and many of these specialised plants fail to set any seeds at all. Such an accident would, of course, seriously affect the numbers of such plants, so that we must not make the mistake that many writers upon this subject have made, that plants specially adapting themselves to cross-fertilisation by insects are more greatly advantaged than the self-fertilised species. Looked at from the human standpoint, such specialisation has produced finer flowers, for it is worthy of note that no member of the Ranunculus family develops blue colouring unless it is specially adapted for cross-fertilisation by bees.

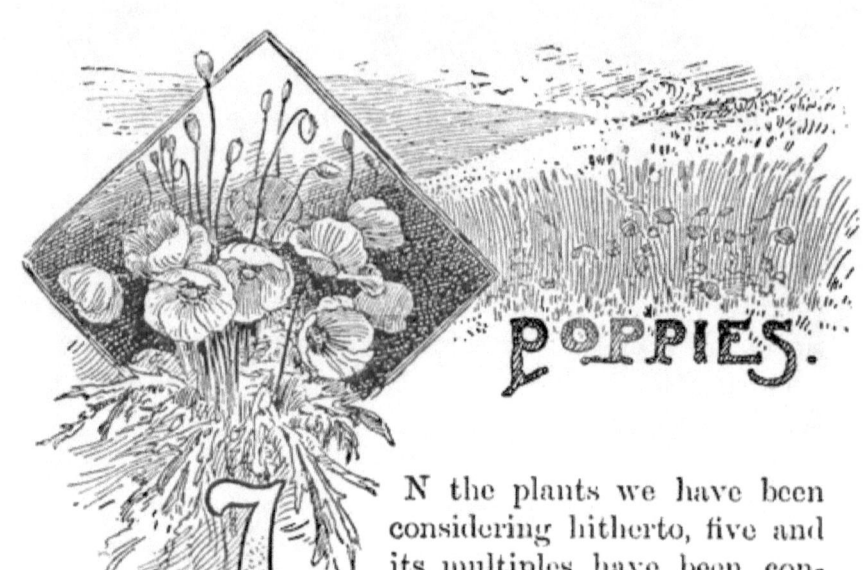

POPPIES.

*I*N the plants we have been considering hitherto, five and its multiples have been conspicuous in the arrangements of the floral whorls. Individual plants or species have shown a tendency to vary to four or a higher number not strictly a multiple of five, and in some of these the reason may be found. But in the Poppy family, which has but few representatives in this country, we find a remarkable uniformity in the number of flower-parts. Two is the distinguishing number: two sepals, four (2+2) petals, stamens and stigmas some multiple of two. The plants are remarkable for an abundance of juice, which readily flows if the stems or leaves be broken. Our native species may be separated into two groups by the colour of this juice, one set having a white milky sap, and in the others it is yellow. Further, the possession of yellow juice is correlated with yellow flowers, and the white juice with scarlet flowers. From their colour and the character of their seed-vessels it is probable that the yellow-flowered species are the more ancient forms of Poppy-worts.

Probably the best known of the yellow species is the Yellow Horned Poppy (*Glaucium luteum*), which is so striking an object in many places on the seacoast, on account of its large solitary flowers, which are three or four inches across. Its leaves, as commonly happens with seaside plants, are thickened, and very handsomely lobed and cut; they are of the pretty bluey-greyish-green colour that distinguishes many plants of the shore, and which is commemorated in the name Glaucium. As the flower-bud begins to expand, the sepals are thrown completely off, and the petals are seen to be crumpled and creased. As soon as the stigmas have received pollen, the petals and stamens also fall off, so that all these organs are said to be *caducous*. The petals form two opposite pairs, of which the inner ones are smaller than the outer. An indefinite number of stamens surrounds the long two-celled ovary, and the seed-eggs are attached to two projections called *placentæ*, which are outgrowths from the inturned edges of the carpels forming the ovary. The two stigmas expand just over these placentæ, and are easily fertilised by pollen

Horned Poppy

from the same flower, for the stamens stand erectly round the pistil, and form a more convenient alighting-stage for flying insects than the soft petals. The fertile ovary lengthens into a seed-vessel called a silique, which attains in this species to the extraordinary length of a foot,—three or four times the diameter of the expanded flower,—and opens from the top by two valves almost as long.

The Greater Celandine, or Swallow-wort (*Chelidonium majus*), is another yellow-flowered Poppy-wort, which grows along hedge-banks and in all sorts of waste corners. Its thin leaves are much divided pinnately, and present a very different appearance from those of the Horned Poppy. The flower-stalks, too, instead of ending in one large flower, bear a cluster of four or five small ones. These are not more than an inch across, each with a slender little foot-stalk of similar length, and as these all radiate from a common centre they form an umbel.

The flowers do not secrete honey, consequently the only insects that visit them are the pollen-seekers, though several of these are humble-bees that ordinarily look for honey also,—such as the Field Humble-bees (*Bombus pratorum* and *B. agrorum*), as well as *B. rajellus* and several species of *Halictus*. These alight upon the stamens and stigma in the centre of the flower, and as the stigma is taller than the anthers, there is considerable chance of cross-fertilisation in all these cases; but with the humble-bees it is certain, because these make straight for the stigmas. The fruit is similar to that of the Horned Poppy, but only an inch and a half in length, one-celled, and the valves open from the bottom upwards.

The Welsh Poppy (*Meconopsis cambrica*) has flowers almost as large as those of *Glaucium*, but of pale sulphury-yellow instead of a deep golden hue. The ovary is one-celled, and the four or six stigmas form a radiating head to the distinct style. The fruit is a more oval capsule, and opens just below the style by as many little valves as there are stigmas. This species, it will be seen, leads up to the true Poppies with red flowers.

The smallest of the red section is the Long Prickly-headed Poppy (*Papaver argemone*), whose petals are small and narrow, so that the flower does not resemble the Horned Poppy in its fulness or cup-shape, but forms a cross of petals with greater space between each than is found even in the Swallow-wort. It grows in dry, waste places, on the tops of banks and walls, and its weak, pale appearance would give one the impression that it is merely a specimen of the Common Poppy of the cornfields, poverty-stricken owing to the seed having fallen on stony ground, but there are several small differences into which we need not enter minutely. Two will serve: the anther filaments in the present species become stouter upwards,

Long Prickly-headed Poppy

and the seed-vessel is club-shaped, bristly, with a somewhat conical roof, on which from four to six stigmatic rays will be seen. If this seed-vessel is cut across, it will be found to be divided into a large number of compartments by the outgrowth of plate-like placentæ from the walls of the capsule, and each of these compartments opens by a little window under the eaves of the stigmatic roof. They are closed until the seeds are ripe, then the little valve shrinks away from the stigma, and as the wind sways the long slender stalk the numerous and minute round seeds are shaken out. There is another species with capsule of similar shape, the Long Smooth-headed Poppy (*P. dubium*), but in this the filaments of the stamens are thin and thread-like throughout their length, the petals are broad, and the capsule is not bristly; moreover, there are from six to twelve stigmatic rays on the roof. Then we have the Round Rough-headed Poppy (*P. hybridum*), with small flowers, but larger than the Prickly Long-head, though its stamen filaments are similar, and the capsule bristly; but the capsule is globular in form, and the rays range from four to eight. The Common Poppy of our cornfields (*P. rhœas*) differs from all the others in the size of its petals, the two outer ones being so large that until the flower is fully expanded they enclose the smaller pair. The short, round capsule is quite smooth, and is further distinguished from the Rough Round-head by its eight to twelve rays, and by being slightly elevated above the receptacle by a little foot-stalk.

Now, all these Poppies, yellow and red, have so many features in common, that no botanist would hesitate to say they are all descended from a common ancestor, and that ancestor belonging to the yellow-blooded race. The differences in the form of the capsule between *G. luteum* and *P. rhœas* is very

Common Poppy.

great, and, coupled with the different-coloured petals and sap, might be held to indicate separate origin; but even in the seven species that alone represent the entire family in our flora, the intermediate stages are so well suggested that no difficulty is raised. Even the importance of the different colours of the juice is minimised by the occurrence of a sub-species of *P. dubium*, whose white juice turns yellow on exposure to the air. The four red Poppies are annuals; the Horned Poppy is usually annual, sometimes biennial; the Welsh Poppy and Greater Celandine are both perennial. None of the species appears to produce honey, insects being attracted by the showy petals and coming for pollen alone; alighting on the convenient stigmas, and easily reaching the anthers, they easily effect self-fertilisation, but no doubt crosses are very frequent. Flowers of this open character are not specialised for fertilisation by any particular species or family of insects; they are visited by beetles, flies, and bees indiscriminately.

WALLFLOWER AND CABBAGE

TO readers of very refined tastes the association of a sweet-scented, beautifully-coloured flower with the coarse, rank-smelling, and altogether vulgar Cabbage will appear perfectly incongruous; but it is Nature that must be blamed if any fault is to be found with such arrangement of my material. For she has put the sweet-smelling Wallflower, Stock, and Rocket into the same family with Cabbage and Turnip, Radish and Horse-radish, and a host of quite mean and uninteresting weeds that the ordinary man would declare to be neither ornamental nor useful. There are, however, many things that help to make this planet habitable, but whose utility the average man fails to see. The weeds that are apparently the most worthless and least ornamental are incessantly providing the enormous volumes of oxygen demanded by animal life, many of them working through that period of the year when "the useful and the ornamental" are leafless; at the same time, they and the insects that feed upon them are making use of solar force and converting it by their vital chemistry into material whereby the soil is enriched for more important plants.

The Wallflower (*Cheiranthus cheiri*) is perfectly well known to all, although, strictly speaking, it is not a native plant, but for more than three centuries it is known to have been growing wild upon our old walls. It is a perennial plant, the lower part of its stems being shrubby and enduring. Its leaves are simple, of a narrow lance-shape without stipules, arising alternately from the stems. Respecting the number of the organs in all this large family of Crossworts (*Cruciferæ*), there has been considerable discussion whether it should be regarded as two, or four, or a reduction from an originally larger number, though four appears to be the more generally accepted number. There are four sepals in two pairs, four petals arranged crosswise, and six stamens— two being held, on the tetramerous

Wallflower.

hypothesis, to have been suppressed. Though the ovary is usually one- or two- celled, it is believed to have had origin in the amalgamation of four carpels possessed by a remote ancestor. It is crowned by a couple of stigma-lobes with a notch-like space between them.

Now let us look at these several parts in the Wallflower, and see if we cannot get a note or two of interest from them, in further illustration of that ready adaptability to surrounding conditions which appears to be shared by all plants in some degree. The sepals of the Wallflower are long and stiff; they stand perfectly erect, and are distinguished as "front," "back," and "side" sepals. There are, of course, two side sepals, and their edges overlap the back and front ones, so that a kind of square tube is formed. The side sepals differ from the others in the fact that their base is swollen into a little pouch. The four petals are arranged diagonally to the sepals, so that they fit into the spaces left between the pointed tips of each two neighbouring sepals. If we strip the sepals from a flower, we at once find that the petals, the moment they get out of sight in the sepal-tube, become exceedingly slender. This is another example of floral economy. The plant has found it pay to adapt itself specially for the visits of long-tongued insects, but instead of developing a long spur to its petals or sepals, like the Columbine and Larkspur, or turning the entire flower into a long slender tube of one piece like Honeysuckle and Convolvulus, it has lengthened and stiffened its sepals without uniting their edges, and has reduced the lower half of its petals to mere threads, the broad, heavy, and showy blades being supported by the tops of the sepals.

If now we strip away a couple of these petals as we have already pulled off the sepals, we shall find four of the stamens to be very long, standing erectly in opposite pairs, whilst the remaining two are much shorter and bend away from the pistil. It will also be seen that these short stamens stand in front of the side sepals with the bulged base, and if we examine these two sepals closely, we shall see that at the base of each there is a little glandular disk which secretes honey, and this flows into the pouched base of the sepal. The two short stamens bending outwards leave a tolerably open space on either side of the pistil down to the honey-pouch.

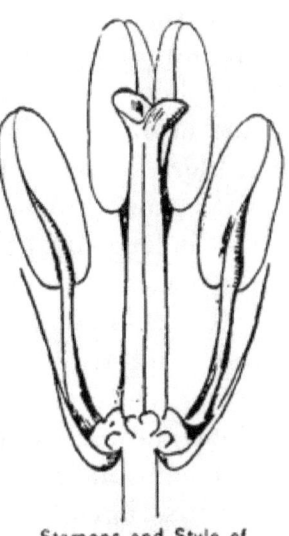

Stamens and Style of Wallflower

A long-tongued insect alighting on the expanded portion (limb) of the petals, must lay its trunk in the stigma notch and pass it between the ovary and the short stamen in order to reach the honey, and in so doing gets it dusted with pollen from the anther. On visiting another flower, this pollen will be scraped off the insect's proboscis by the stigmas as it passes between them, and so cross-fertilisation be brought about. The ovary lengthens into a long slender pod like that of the Horned Poppy, but straight and much shorter. It sets the seeds free by the splitting off of the valve from bottom to top.

Seed-vessel of Wallflower

The Hoary Stock (*Matthiola incana*), from which the Brompton Stocks of the garden are descended, and the Great Sea-stock (*M. sinuata*), are very similar in their arrangements to the Wallflower; but they are well-nigh extinct in this country now, and hardly likely to come under the notice of our readers. The first-named has purple or violet flowers, with honey pouches; the Sea-stock is pale-violet, and is fragrant at night,—an indication that it caters for nocturnal moths.

There is not a great variety in the flower structure and adaptations in the Crosswort family, the majority being white or yellow, and self-fertilising. Some of them show clearly that they have fallen back from a condition when they sought insect aid in the fertilisation of their seed-eggs. There is one rough and ready way of telling at a glance whether a Crosswort desires to be cross- or self- fertilised; if the stigma is lobed, as we saw it in Wallflower, it is of the former set; if it prefers to be fertilised by its own pollen, the stigma will form a rounded head. There is no need for the notched stigma to scrape pollen off the tongues of insects that do not come and are not even invited; for correlated with this globular stigma is the absence or reduction of honey, honey-glands, scent, and conspicuous petals. The anthers and stigmas also come to maturity together. Most of such flowers are white and small, but they contrive to produce a large number of seeds, and are everywhere abundant.

Many of these little Crossworts have gone with colonising Europeans almost everywhere, and in many places have become an intolerable nuisance, so rapidly have they found out that no allied species had grown there before to rob the soil of the particular ingredients they chiefly required. About thirty years ago, the Water-cress (*Nasturtium officinale*) having got itself "transported" to New Zealand, had become the cause of much expensive mischief to the rivers and waterways, the stem growing to a length of twelve feet, and as thick as a man's finger, forming network across rivers that put a stop to navigation, except where hundreds of pounds per annum were spent in keeping down the growth. The petals of this plant are white, and larger than the sepals, but still very small,—less than a fourth of an inch across. It is chiefly fertilised by beetles. Its numbers are kept down in this country by the demands upon the plant for table use as an anti-scorbutic.

It is worthy of note that most or all the numerous members of the family share this anti-scorbutic property, which has caused many to be selected for cultivation—such as Cabbage, Turnip, Radish, Mustard, etc. They are rich in nitrogen and sulphur, and it is the giving off of these substances that produces the unpleasant odours from cabbage-fields and the water that "greens" have been boiled in. From the same principles they elaborate the pungent, hot, or acrid flavours which cause many of them to be esteemed by man, though this is a protection set up by the plant to ward off the attacks of herbivorous animals. These plants in the wild condition are but little affected by insects, yet in a cultivated state they are much esteemed as caterpillar food, as a glance at the riddled leaves of our cabbages will certify. This difference is probably due to an attenuation of the pungent principle with the increased bulk of the plant.

The well-known Lady's Smock (*Cardamine pratensis*), which makes bright many a damp meadow with its abundant pale-lilac flowers, is mainly fertilised by insects, but if these do not come in time, the longer stamens press against the stigma and pollinate it. Many other of the Crossworts are self-fertile in the same way. This plant is sometimes found so overflowing with vital energy, that it develops roots from the joints of its pretty pinnate leaves that lie on the damp ground, and then new little plants arise from the axils of the leaflets. If the flowers of such specimens be examined, they will be found to resemble little double roses, for the ovary in each has developed into another flower, with calyx and corolla complete, and probably yet another flower growing out of that.

The five native species of Hedge Mustards (*Sisymbrium*), so common under every hedge, are habitually fertilised with their own pollen by beetles and flies chiefly. Some of these have small yellow flowers, others white, arranged in a loose raceme,—that is, each on a short foot-stalk attached to the main flower-stem. Among these the yellow-flowered London Rocket (*S. irio*) is noteworthy, as having sprung up suddenly after the Great Fire of London in 1666, wherever the ground was left bare by the ravages of the fire; it owes its name to this circumstance. Jack-by-the-hedge, or Garlic Mustard (*S. alliaria*), and Hedge Mustard (*S. officinale*), secrete honey, and are visited by flies, beetles, and an occasional bee or butterfly, and these must effect crosses.

The Wild Cabbage (*Brassica oleracea*) is now found only on the sea-cliffs of our south-western shores, forming a stout crooked stem a couple of feet in height, strongly scarred where the leaves have fallen. The long lobed leaves have that glaucous "bloom" that distinguishes some of the cultivated varieties. The pale-yellow flowers are about an inch across, with erect sepals like those of the Wallflower. It is wonderful to think that people can deny to Nature the power of producing, during ages, many forms of animals or plants from one ancestral type by merely changing the conditions under which they live; whilst within the historical period man has produced from the Wild Cabbage such dissimilar table vegetables as Scotch Kail, Borecole, Savoys, Brussels Sprouts, Red, White, and Cow Cabbages, Cauliflower and Broccoli. All owe their origin to the very different-looking Wild Cabbage of our sea-cliffs. There are not wanting those who believe that in addition to the cultivated varieties named, the Rape, the Turnip, and the Swede are also referable to *B. oleracea* as the common ancestor.

The small but ubiquitous Shepherd's Purse (*Capsella bursa-pastoris*) may be taken as a type of the inconspicuous-flowered weeds of this family, which fertilise themselves, and produce such abundance of seeds that they take possession of all cultivated ground so soon as the husbandman's back is turned. Their flowers range in diameter from one-fourth to one-twelfth of an inch, and in some cases the petals have been converted into stamens as being more useful to plants once dependent upon the visits of insects, but which have now learned to do without them. The presence of these minute white petals, and in some cases honey-glands that no longer secrete honey, testifies to the fact that these plants have come down in the world. Yet, in spite of their lack of show or " presence," they are a standing rebuke to those writers who have so strongly asserted that cross-fertilisation produces a more vigorous and successful race. What cross-fertilisation by insect agency does is to produce more brilliant individuals, and to keep up large flowers of bright hue. In fact, it produces a kind of floral aristocracy; whilst the principal work of the vegetable kingdom — the abstraction of carbon from the atmosphere, the setting free of oxygen, the production of food for the entire animal races—is done mainly by the less brilliant weeds and grasses and trees,—the working classes.

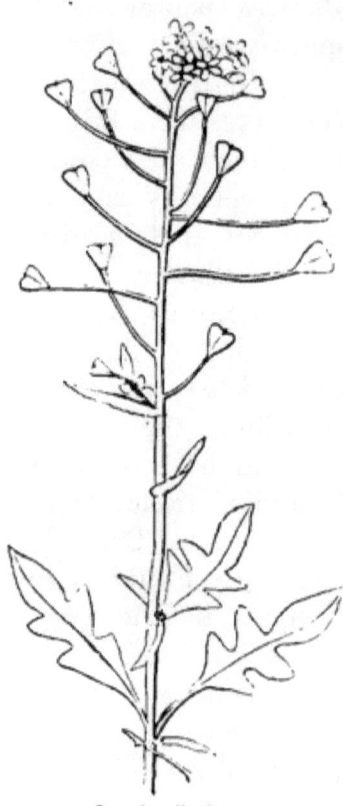

Shepherd's Purse

Sir Joseph Hooker has recorded his opinion that, of the numerous weeds of this character that cling to the skirts of husbandry and seldom appear on virgin soil, the Shepherd's Purse would be among the first to disappear entirely if the soil were unleft disturbed by man and the animals he rears. It has long been

observed to be a mere hanger-on of industry, and it
has earned the alternative title of Pickpocket, from a
well-grounded suspicion that it robs the farmer of
much of the plant-food he has distributed over his
fields for his crops. The Shepherd's Purse owes this
name to a striking departure in the form of its
seed-vessels from those of the Wallflower and many
other Crossworts; it is somewhat heart-shaped, and
resembles an old-fashioned form of purse—the two
valves, which open from the top, being somewhat
boat-shaped. The plant may be only a few inches
high, or it may grow to a couple of feet. The flower
is only about one-tenth of an inch across, having no
honey, no scent, and being capable of fertilising itself
without insect aid, though flies visit it for the sake
of its pollen.

The Annual Candytuft (*Iberis amara*), which is
one of several species cultivated in flower-gardens,
occurs wild in central and eastern England, but with
much smaller flowers. Its white or lilac flowers grow
at first in flat heads (*corymbs*), but ultimately, by the
continued lengthening of the growing point, these
become racemes; they are noteworthy, however, for
the fact that the two petals on that side of the flower
that is turned towards the centre of the corymb are
smaller than the two that are turned towards the
circumference. This is more striking in the outer
row of flowers, and we shall find other instances of
plants in widely separate families adopting a similar
method of making their flower-clusters more attractive
to insects. The class of insects that frequent Wild
Candytuft has not been recorded, but the notched
stigma and these enlarged petals are presumptive

evidence that it is insect-fertilised, even apart from the popularity of the cultivated Candytuft with insects.

Although Woad (*Isatis tinctoria*) cannot claim our attention on account of its yellow flowers, which are very tiny, it is yet deserving a word because it has played an important part in our history. From our earliest lessons in that branch of knowledge, we have all been made acquainted that the ancient Britons were principally clothed in a more or less artistic design in blue. Cæsar has recorded the fact that on this account they were designated *brit* or *brith* (whence Briton), which is Celtic for painted, and he further tells us that this pleasing stain was obtained from the juice of Woad. Well, that is a fact that should entitle an insignificant plant to respect, and it is not surprising to learn that the dyeing industry still finds employment for Woad in its operations, though it is as a mordant rather than an actual dye it is used to-day. The seed-pods differ in form from those already described; they have a thick wing all round the seed-capsule, and assume a pendulous attitude as soon as the petals have fallen.

VIOLETS AND PANSIES

THESE beautiful flowers have been celebrated in poetry and romance from very ancient times, and they have still considerable vogue not merely with sentimental folk, but also with the scientific botanists who are not usually reckoned in that class. Every bit of the Violet-flower is crowded with interest. It is a flower that has been most carefully and completely adapted for cross-fertilisation by insects and to prevent self-fertilisation, yet strange to say, these flowers rarely produce seed in this country. To me this has always been a puzzle. On all the hypotheses respecting cross-fertilisation this thing is a paradox. It can only be explained on the assumption that the insects which successfully fertilise it in the warmer parts of the Continent do not occur in Britain, though they must have been indigenous until quite recent times, or the flower would have degenerated. But there are no signs of degeneration about our Sweet Violet (*Viola odorata*), though it has devised a very ingenious way of getting over the difficulty thus presented to the perpetuation of the species.

First let us refresh the memory of the reader respecting the structure of the flower he—in common with civilised man generally—knows so well. The short stem or rootstock, with its runners and heart-shaped and stipulate leaves, needs no further mention, but the flowers we see at once are pentamerous—or in whorls of fives—five sepals, five petals, five stamens. Yet the flower has not the regular form observable in the Roses and Buttercups, where all the petals were of one form and size, neither is it so irregular as the Monkshood. The sepals are almost equal, but they have each a singular flap-like growth backwards beyond their attachment to the receptacle, and they remain attached to the seed-capsule when this is developed. The petals are irregular in size and shape, there being two pairs and a larger odd one that is developed backwards as a short, slightly curved spur. As the flower grows, this odd petal is the lowest of the five, though it is really the upper petal, but to suit the convenience of bee-friends the plant has taken to bending over the upper part of the flower-stalk, so that this petal should come lowest and serve for a platform. This large petal is also marked with a number of dark hair-streaks *which all converge to the entrance to the spur.* That these are of similar purpose to the direction-posts we set up at cross-roads, and intended to guide insects to the store

Section of Violet

of honey there can be no doubt, for they only occur in flowers that are specially adapted for insect-fertilisation. Another point worth noting is that in regular flowers, like Pinks and Wood Sorrel, these guides appear on all the petals, but in irregular flowers they will be found only on the platform petal, and perhaps on the one at each side of it, as in the Pansy.

The stamens are irregular to this extent, that two of them are prolonged backwards into honey-producing spurs and lie within the hollow spur of the large petal. The entire five lie around the pistil, and the anthers are so broad that their edges fit closely together, whilst at their tips are broad expansions which are pressed against the curved style. They open on their inner faces, and discharge their pollen into the chamber formed by the contact of their edges. This pollen differs from that produced by the majority of entomophilous flowers in the fact that it is dry instead of being sticky. Owing to the manner in which the Violet-flower is hung on its stalk, the enlarged head of the style hangs down with the stigmatic surface below, so that if a bee alighted upon the broad petal and followed the guide-lines towards the honey, its head would come in collision with the stigma, and any pollen it might have upon its face would adhere to the stigma. It may be objected that a bee's face is not the usual place to find pollen-grains, but if it had previously visited Violets it would be so adorned. Take a Violet and a bristle or fine grass-bent; hold the Violet in the natural position, and pass the bristle inside the spur, moving it as the tongue of a bee might search for the honey-glands. On pressing against the stamen spurs the leverage has the effect of dislocating the ring of anthers, and the dry pollen will fall out upon your fingers, which stand in place of the bee's face. The bee's head pressed against the stigma will effect the same thing, owing to the curved style. If you will at once repeat the experiment with another Violet, you will find that some of the pollen-grains from your finger-tips adhere to the stigma!

The Sweet Violet certainly does all she can to take advantage of this highly specialised flower, but it must be admitted that her efforts generally end in failure. The poets and moralists have always extolled the Violet on account of her modesty,

> "Like virtue oft unseen, unknown,
> Save by the sweetness, round it thrown;"

but they ought to add patience also to her attributes, for she perfects these wonderful arrangements in each flower, ever hoping to woo the bee, and patiently waits, yet he seldom comes. Not one in a hundred of such flowers produces a single seed, in spite of royal purple hues, fragrance, and honey. The probable reason is that the flowers are produced at too early a season; certainly there are few bees about in early spring, and possibly such things as Sallow-bloom ("Palm") then offers greater attraction to those that are about. But the Violet is a resourceful plant, and she has not exhausted either her art or her means, though her purple flowers have been a considerable drain. She enlarges her leaves in the hope of retrieving her losses by additional production of nutritive material, and sends out rooting runners. From the joints of these runners flower-buds are produced, but they never expand; sometimes there are minute green petals, but more often no petals at all. There is no honey, no scent, no colour other than green, but there is an ovary and stigma, with a couple of stamens producing only a little pollen. The anthers lie over the stigmas, and the pollen-shoots actually penetrate it without leaving the anther, and as the ovary develops, the stamens are dragged from their former attachment and carried up with it. Every seed-egg in the ovary appears to be fertilised in this way, and in due time you may find abundant seed-capsules upon the Violet-plant that the bees ignored, and you may watch the clever way in which she distributes her seeds to a distance, so that the seedlings may not overcrowd her, or one another.

The seed-vessel splits into three valves corresponding with the three carpels of which it was composed, and these lie far apart, with the glossy seeds in the middle of each, in order that they may become harder. Then each valve begins to fold as though the middle were hinged, and the two edges are brought near together.

Full of seeds   Seeds discharged

Violet Capsule

The pressure is increased until one after the other the hard shiny seeds are shot out with great force to a distance of several yards, where their smoothness enables them to sink readily between the grass or moss to the moist earth beneath.

We have seven British species of Viola, and the arrangements for cross-fertilisation are very similar in each, but the Pansy (*Viola tricolor*) appears to be the only one that gets any considerable benefit out of its perfect flowers; and consequently *it does not produce these imperfect* (cleistogamous) *flowers!* The Marsh Violet (*V. palustris*), the Hairy Violet (*V. hirta*), the Dog Violet (*V. canina*), the Wood Violet (*V. sylvatica*), and the rare Sand Violet (*V. arenaria*) differ only in comparatively small points, which we need not discuss. They have all purple or blue flowers, but the Wild Pansy differs in more striking fashion from the others.

There is no suggestion of woodiness about its rootstock; its stems are long and angular; its stipules are leafy and copious; its leaves, instead of the general heart-shape of the Violets, are more lance-shaped, with large rounded teeth; the sepals are much longer—in some forms longer than the petals—they have larger "ears," and the two uppermost petals instead of leaning forward and slightly curling at their edges as do the Violets, keep quite upright and flat.

Ovary and Stigma of Wild Pansy

The colour of the petals is not uniform as a rule, but it differs in individuals. Usually purple, yellow, and white are combined in the same flower, but in varying proportions. An important difference is seen in the character of the style and stigma. The style is short and straight, but has a curved base which constitutes a kind of spring, always keeping the stigma pressed close to the large platform petal. The stigma itself is a large round head, having a mouth-like orifice near the front with a prominent lip to it, and a tuft of hairs on either side.

The meaning of these departures from the simple obliquely-cut thickened end of the style in Violets is seen if again we imitate an insect's visit by means of the flower and bristle. The opening to the spur is so restricted by a fleshy growth from each of the intermediate petals that the only way to the honey is beneath the skull-like stigma. If the insect's proboscis has already been dipped into a Pansy spur and has come away with pollen attached, *that pollen will now be scraped off by the sensitive lip of the stigma*.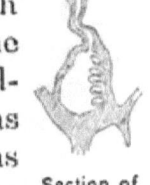

Section of Pansy Ovary

On the other hand, to make as certain as evidence can testify that this is an arrangement to favour cross-fertilisation and prohibit self-fertilisation, when a bee's tongue has got dusted with pollen inside this spur and is withdrawn, the slight pressure effectually shuts up the lip and closes the sensitive cavity against its own pollen. These flowers *do* produce plenty of seeds, so there is no necessity for cleistogamous flowers in this species. But there is a varietal form or sub-species with petals usually shorter than the sepals or wanting altogether, and the stigmatic cavity more to the side of the "skull," and without the lip. Pollen-grains fall spontaneously out of the anther cavity before or just after the opening of the flower and into the unprotected cavity. Müller, who first called attention to this remarkable contrast in the two forms, says:—

"When the visits of insects are prevented by a fine net, the flowers of the small-flowered form wither two or three days after opening, every one setting a vigorous seed-capsule; those of the large-flowered form remain in full freshness more than two or three weeks, at length withering without having set any capsule; when fertilised, they too wither after two or three days."

These two forms of flowers are not found upon the same plant, and in all probability they indicate the evolution of a new species, though at present the general dissimilarity of the two forms is not sufficiently marked to separate them.

Pinks and Chickweed

WE have seen in the great Rose family the extremes of small trees covered with fine blossoms on the one hand, and inconspicuous degenerate weeds on the other; in the Buttercup family, the simple open yellow flowers of the typical group and the highly specialised blossoms of Monkshood and Larkspur; and now we come to consider a family ranging from the highly-developed Pinks to the degenerate Chickweeds and Pearl-worts. The Pink family is characterised by its slender stems being jointed, the ends of the joints (*nodes*) being much thicker than the intermediate portion (*internode*). At the joint the leaves are given off in pairs, and their bases are usually so enlarged that they surround the stem and are joined together (*connate*). These leaves are always what is known as entire: they are never broken up into lobes or leaflets, and they rarely have stipules: if these are present, they will be very small, and of a dry, skinny texture.

The general character of the flowers of this family may be readily understood by calling to mind or examining a (single) Pink or Sweet William from the garden, but we would rather our readers referred to the wild plants from hedge or wood, and we may get a specimen of Red Campion from the moist banks of a deep, sheltered lane almost any day in the year. If we pull this flower to pieces, we can scarcely fail to be reminded in some way of the flowers of the Crosswort family. The calyx is stiff and upright, and the petals, though broad and spreading at the top of the calyx, are exceedingly slender within. But there is a marked difference—the five sepals have *got their edges joined together*, so that the calyx has become a downy, reddish tube, with five triangular teeth at its upper end to show that it really consists of the five united sepals. The five petals are separate, rosy-pink in colour, and the limb or broad portion is deeply cleft into two lobes. At the base of each limb there are two little scales of a paler tint standing up, so that the ten form a kind of coronet at the top of the corolla-tube. There are ten erect stamens, and as a rule the flower that contains these has no pistil; whilst the flower bearing the perfect pistil and five styles contains no stamens. Further, as a rule, a plant will produce complete male or complete female flowers only. This separation of the sexes by being borne upon different plants does not characterise the family, but is peculiar to this species—a condition known as diœcism. Cross-fertilisation is, of course, unavoidable. The seed-vessel is somewhat egg-shaped, with a tendency to be globular, and opens at the top by splitting into ten teeth.

Red Campion.

Now, there is another species of Lychnis very similar to this, so far as form goes, and no doubt it was once a mere variety of it. This is known as the White Campion (*Lychnis vespertina*), and as the name indicates, its flowers are entirely white; as usual with large white flowers, they open properly only in the evening, and then give out a fragrant odour to attract crepuscular moths. On a moonless night these flowers positively gleam along the hedgerow, and as they are usually backed by dark vegetation they must be very distinct to night-flying insects.

Occasionally a Red Campion appears with white flowers, or a White Campion with red, and at first sight this fact might be taken to indicate that the two are mere varieties of one form; but there are other differences—for instance, the calyx of White Campion is green, with longer teeth, the seed-vessel is more conical, and the teeth by which it opens are shorter and remain erect, whilst those of Red Campion curl outwardly, and so open the mouth of the capsule more widely. Then there is the different hour for opening and the fragrance. None of these points is of any great importance perhaps by itself, but the sum of them witnesses that the white variety has been so long separated from the red that the differences have become permanent and are transmitted to successive generations.

The appearance of white flowers (*albinism*) on the Red Campion is what we expect to find in the case of most red or blue flowers; it shows how White Campion originated. On the other hand, the occurrence of reddish flowers on White Campion is a reversion (*atavism*) to an earlier form, and also indicates whence the white species has come. We can go back a step farther, and trace the Red Campion to the Ragged Robin (*L. flos-cuculi*), which grows in bogs and wet meadows with Yellow Iris and Cotton-grass. In all essentials this is similar to Red Campion, but its rosy petals are deeply cleft into four long slender lobes which give the ragged appearance indicated in the popular name. In addition to the difference in the shape of the petals, we find that each flower contains both stamens and pistil, and the seed-vessel opens with five teeth only. Ragged Robin secures cross-fertilisation by bringing its anthers to maturity and shedding its pollen before the stigmas are ready to receive it. The honey induces insect-visitors to act as pollen-carriers from flower to flower, and from plant to plant. Bees, butterflies, and long-tongued flies are the carriers, and they cannot get at the honey without dusting the proboscis with pollen in one flower and transferring it to the stigmas of the next.

The genus Silene ("Catchflies") differs from Lychnis in the fact that there are usually only three styles, but the general structure and habits of the species are the same, and point to the two genera having a common ancestor not far back in the evolution of the race. One species, the Bladder Campion, or White-bottle (*Silene cucubalus*), resembles the White Campion at a little distance, but its calyx is greatly distended, and of a smooth parchmenty character that quite justifies the title of Bladder. The flowers are of three kinds: (1) with stamens only, (2) with pistil only, (3) with both stamens and pistil.

What is the purpose of the bladder-like calyx it is not easy to see; but an enemy of the plant makes use of it for his own protection, and unless it served some useful purpose to the plant, we should expect natural selection to do away with it. There are several species of Noctuid moths whose caterpillars feed on the unripe seeds of certain members of the Pink family. One of these is called the Tawny Shears (*Dianthœcia carpophaga*), and another the Campion-moth (*D. cucubali*). Their eggs are laid in the flowers in June, and about a week later the young caterpillar is hatched, bores into the growing seed-vessel, and begins to eat the unripe seeds. Having demolished these, it passes to an older capsule, and if too large to enter it, lies coiled in a ring round it within the calyx-bladder, so that this becomes for a time a protection for the plant's chief enemy—one that partially prevents the plant from perpetuating its kind. Finally the caterpillar gets too large to be thus accommodated, and it is compelled to hide at the roots of the plant during the day to escape bird-enemies. In the evening it crawls up to the seed-vessels again, where it may be found with its head and fore-parts hidden in the calyx, and its hinder part exposed.

Sea Campion

There is another plant, called the Sea Campion

(*S. maritima*), differing from the Bladder Campion in such minute and obscure points that only a botanist would appreciate them—and some botanists believe the one to be a mere variety of the other. A very similar moth—the Pod-lover (*D. capsophila*)—feeds upon it in the caterpillar state. Now, these insects differ chiefly, in both caterpillar and winged stages, in the matter of colour; so there are not wanting entomologists who regard them as mere variations of the same species. Here we have, then, a very close and interesting parallel between the cases of the plants and the insects. It should be added that the moths

Sea Campion
Male flower

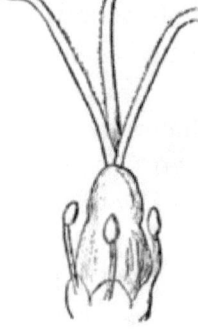

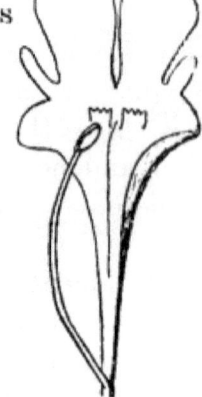

Female flower   Female flower enlarged   Petal and Stamen
Sea Campion

mentioned assiduously fertilise the flowers, and so insure an abundance of seed-vessels for their offspring.

The typical genus of the family is Dianthus, which includes the Pinks, Carnation, Sweet William, etc. Most of the six species admitted to our floras are either very rare or doubtful natives; the Maiden Pink (*D. deltoides*), however, is a genuine native, of

Sea Campion.

fairly wide distribution in dry soils, and may therefore be taken as the best type available. Our gardens will always furnish examples in the Carnation and Pink, if the Maiden Pink be not available. Like the garden forms, the wild species has thick, grass-shaped leaves, a long tubular calyx with five teeth, five rosy petals with toothed edges, ten stamens, and two styles. The petals spread away from each other, and provide an admirable alighting-platform for insects, though there is no fragrance to attract them, nor can any except butterflies and moths reach the honey that is secreted at the bottom of the long tube, where stamens and petals join. The opening to this tube is so narrow that smaller insects do not attempt to get in, and as the stamens mature five at a time, they erect themselves, and the anthers protrude from the tube and discharge their pollen outside. As the first five are withering, the second set take their places, and discharge their pollen. The moths or butterflies that come for honey are almost sure to get the pollen from one or other set of anthers attached to their hairy heads, and to carry it to other flowers. When the second five anthers have shed all their pollen, the two styles begin to lengthen, and continue to do so until the stigmas are far outside the tube, when they curl away from each other so as to occupy the

Maiden Pink

space formerly covered by the anthers. A pollen-dusted insect from a younger flower cannot fail to fertilise these stigmas. Self-fertilisation is out of the question here.

The largest of our native species is the Corn Cockle (*Githago segetum*), a beautiful annual weed of corn-fields, which agrees with the Pinks in having honey accessible only to lepidoptera; but in its structure it comes nearer to the Lychnis, though it has no scales on its petals. The plant is densely covered with long white hairs, calculated to prevent small crawling insects, ants and the like, from climbing up to the flowers and stealing the honey and pollen. The strongly-ribbed leathery calyx ends in five teeth, which are drawn out into woolly, leaf-like extensions, much longer than the purple petals which are marked with guide-lines. As these flowers are so large and conspicuously coloured, they occur singly on long flower-stalks, and as the anthers shed their pollen before the stigmas are mature, cross-fertilisation is here again imperative.

I have already alluded to the risks run by plants whose flowers are so fitted to prevent self-fertilisation; but no doubt the *species* is always taken care of by a few flowers varying from the prevailing conditions. Thus, if the stigmas of a few flowers matured very quickly after the anthers shed the pollen—that is, before it could all be carried away by insects—self-fertilisation would certainly be effected. There are a number of species in this family that bear indications of having once been cross-fertilised, but evidently finding that such a condition was not the best suited to their particular mode of life have abandoned it in favour of self-fertilisation. And with that abandonment has gone the conspicuous petals, which have dwindled in size though still bearing evidences of their former glory. Thus in the genus Cerastium, or Mouse-ear Chickweeds, some species are cross- and some self- fertilised, yet all contain a disk of five honey-secreting glands, all have white flowers. Though five is the proper number of the flower-parts, as throughout the family, yet in the Erect Mouse-ear (*C. quaternellum*) they have been reduced to four, and the entire plant is smooth, whereas most of the other Mouse-ears are covered with hairs or down, often sticky, in order to discourage ants and other crawling flower-robbers. Now this Erect Mouse-ear does not care any longer if the ants do invade its flowers and kick the pollen from the anthers to the branched stigmas. It has therefore abandoned the hairiness still retained by its first cousins, it has become annual instead of perennial, and it has allowed its petals to become smaller than its sepals.

The Little Mouse-ear (*C. semidecandrum*), that we find blossoming in early spring on every old wall, has petals not more than half the length of the sepals; and although the Broad-leaved Mouse-ear (*C. glomeratum*) has kept its petals as large as its sepals, yet it frequently refuses to open its buds, and when this is done forcibly by the inquiring botanist he finds the anthers have discharged their pollen directly on the stigmas. This species has evidently determined to do without insects altogether in the near future, for sometimes it may be found to have no petals at all.

The Field Mouse-ear (*C. arvense*) is a perennial,

and has retained its liking for cross-fertilisation, therefore its five petals are pure white, twice the length of the sepals, and divided at the tip to give greater prominence to the flowers; but in the absence of insects it fertilises itself. These differences within the limits of a small genus are very instructive, and in the next genus we meet with very similar examples.

The Chickweeds, or Stitchworts (*Stellaria*), include several well-known species, the most conspicuous being the Greater Stitchwort (*S. holostea*), which always gives joy to the flower-lover in spring who comes across its large pure white blossoms in the hedgerow, as represented in the plate. Until the flowers open it might pass with the ordinary rambler as a grass, so slender are its glaucous leaves and brittle jointed stems that have to lean against other plants for support. It has a perennial rootstock, and it holds its place in these crowded situations by having adopted the grass-like form of leaves and stems, which enable it to fill up gaps where plants with broader, more spreading leaves would be starved. The petals are divided at the end into two lobes similar to those of the Campions, but the sepals are quite separate, instead of being united into a strong tube. There are ten stamens, and these mature in two sets of five, just as we saw in the Pinks (*Dianthus*): the first five erect themselves until they occupy the centre of the flower, and as they wither the second set succeeds them. Just before all their pollen is shed, the stigmas mature, curl over towards the anthers, and should no butterfly or moth come along with pollen from another flower, self-fertilisa-

Stitchwort.

tion will take place. The honey is secreted by yellow glands at the base of the outer five stamens. Whether any of the seven British species of Stellaria are or are not in the direct line of evolution leading to the Pinks and Campions, a consideration of their forms and structure will show that they are not far off—at most a collateral branch, but more probably in the direct line. The Greater Stitchwort occasionally calls attention to this probability by appearing with its petals jagged at the edges like those of the Pink, instead of with the usual neat and simple division; sometimes, too, one row of stamens becomes converted into petals, and the flower is "double," as in the cultivated Pinks.

The very similar Lesser Stitchwort (*S. graminea*) looks as though it were a starved form of the Greater Stitchwort, due to the drier situations it affects; but it has several distinguishing features. Here we see the beginning of the tendency to join the sepals into a cylindrical tube; these are united at their base only in this species, nevertheless they form a short conical tube. This indicates that the Lesser has possibly been once more Pink-like than the Greater Stitchwort, though the petals are now actually as short as the sepals, for there would scarcely be the tendency to form a tube except in a conspicuous insect-fertilised flower. Two other species share this tendency to a tubular calyx, and of these the Marsh Stitchwort (*S. palustris*) has the petals larger than the sepals, and the Bog Stitchwort (*S. uliginosa*) has them smaller.

The very familiar Chickweed (*S. media*) is probably nearer the original type than those we have named, for it has oval leaves, broader in proportion to their length; but this has evidently been a more conspicuous plant in its time. It is such a common thing for both animals and plants that have "come down in the world," so to speak, to still retain some small evidences of their having "seen better days" that every obscure detail of their structure has interest, though we cannot always catch the significance at a glance. Take the number of stamens: this will be found to be high in plants so highly specialised that they have entirely given up all power of self-fertilisation and are absolutely dependent upon the visits of insects. Throughout the entire family we find that ten stamens is the prevailing number, and we may find the full complement in the Chickweed, though its notched white petals have become smaller than the sepals. It still possesses honey-glands, which attract many insects in early spring, and I have no doubt that it once had flowers as large as those of the Stitchworts (which we have seen retain the power of self-fertilisation), but owing to some obscure cause a certain colony got neglected by insects and had to depend almost entirely upon self-fertilisation for the production of seed. The lack of stimulus given by insect-visits led to the reduction of the petals to a variable extent, and of the stamens to correspond; the object probably being to utilise the material in the production of a larger number of flowers and seeds. Anyway, we find that Chickweed has become an annual, that in winter its petals are often entirely absent, its stamens reduced to three, and to make sure

Chickweed Flower

of the small quantity of pollen being effective in this case it refuses to open its buds properly, but remains with the bursting anthers pressed against the stigmas, and fertilises every seed-egg.

Thus the Chickweed may be said to have originated as a species *by descent* from a more showy perennial member of the same family, which we probably regard to-day as an entirely distinct species. It may be included by theologians among the plants with which the earth is cursed on account of original sin, and evolutionists of narrow views may point to it as evidence of the evil results of self-fertilisation; but the naturalist who takes broad views can only regard it as an instance of the meekness that inherits the earth. It is a plant that—like most genuine weeds—has stooped to conquer. Its production of seed—the true test of floral success, apart from size or showiness—is so sure that it is found in almost every place where civilised man has been. A very striking story is told by Sir J. D. Hooker which—although I have had occasion to quote it in a former work—will bear reproduction. He says: "Upon one occasion landing on a small uninhabited island nearly at the Antipodes, the first evidence I met with of its having been previously visited by man was the English Chickweed; and this I traced to a mound that marked the grave of a British sailor, and that was covered with the plant, doubtless the offspring of seed that had adhered to the spade or mattock with which the grave had been dug."

Considerations of space prevent our dealing at length with the remaining genera of the Pink family, which includes the Sandworts (*Arenaria*), with small open white or pink flowers, in most cases fertilised by insect aid with their own pollen; but the Three-nerved Sandwort (*A. trinervia*) matures its stigmas before its anthers shed the pollen, and the Sea Purslane (*A. peploides*) has the stamens and the pistils in separate flowers on the same plant. The Pearlworts (*Sagina*), Spurrey (*Spergula*), and Sandwort-Spurrey (*Spergularia*) are similarly small and self-fertile. There is a general Chickweedy character in the flowers of these genera, and like Chickweed they are probably degenerate Pinks and Campions whose glory has departed since they flouted the butterflies and moths and bees. The bulk of the large-flowered members of the family may be said to be moth-fertilised, and this accounts for the prevailing white-tinted flowers.

An interesting habit in many of the smaller members of the family consists in the at first erect flower-stalk (*pedicel*) drooping when fertilisation is effected, so that its faded appearance shall not take off the attractiveness of the flower-cluster, and its gradual erection again when the seeds are nearly ripe, by which time the other flowers in the group have set their seeds. This drooping of the fertilised flowers may be seen in the photographic plate of Stitchwort.

Several species of Silene and Lychnis have the calyx and upper part of the flower-stems covered with sticky hairs, to which small flies, aphides, and other insects become glued, and so the plants have been termed Catchflies. It has been ascertained that, in some species at least, the hairs which secrete the viscid matter have also the power of absorbing and digesting the nitrogenous matter derived from the decay of such captives. It is probable that the glandular hairs were at first evolved for the purpose of entrapping and making examples of the insect-thieves that came to steal honey and pollen; but that in order to turn waste substances to profitable use, they developed into absorbent organs, and utilised dead bodies that had formerly been wasted.

MALLOWS

THE paltry half-dozen of species which represents the large and beautiful family of Mallows in this country form an interesting group in various ways: the native species have long enjoyed high reputation in rustic medicine, and the exotics yield valuable fibres, cotton, and our garden Hollyhocks. Yet it is not with these things, but rather with the beauty and arrangements of the flowers, that we are at present concerned. In general the leaves are as broad as long, variously lobed, angled and toothed; they are alternate in their arrangement on the stem, and start life with stipules, but these are thrown off at an early date. The calyx is partially cut into five pieces—or more correctly, the five sepals are joined except at their upper parts,—but there is a second or false calyx (*epicalyx*) formed below the genuine one by three or more little bracts, and it is on the character of these that the divisions of the family depend.

There are five petals which adhere at their base to

a long tube formed by the union of filaments of the numerous stamens. In all the plants to which we have referred in previous chapters the anthers were two-celled; but in the Mallows they possess but one cell. The base of the staminal tube surrounds the many-celled ovary, each cell with a single seed-egg, and the thread-like styles pass up the tube and stand above the anthers, their inner surfaces being the stigmas. Everybody knows the singular form of the Mallow fruits, which in their juvenile days they used to know as "cheeses," from the resemblance to the flattened, round form of certain makes. When ripe, these "cheeses" split into about a dozen segments, each corresponding to a cell of the ovary, and containing a curved seed.

The species best known by name is the Marsh Mallow (*Althæa officinalis*), but it is by no means so familiar to sight as the Common Mallow or the Dwarf Mallow, both of which are popularly known as Marsh Mallows. The Marsh Mallow is covered with soft down, has thick, roundish leaves, and rosy flowers as much as a couple of inches across. But it only occurs locally in marshy places not far from the sea. The Hairy Marsh Mallow (*A. hirsuta*) is a smaller species of much rarer occurrence, covered with stiff hairs, the leaves more kidney-shaped, and the smaller flowers purplish in hue.

The Common Mallow (*Malva sylvestris*), standing in the shelter of a meadow-hedge just about the time the hay is cut, is a very beautiful plant; growing to a height of three or four feet, well covered with the large purplish flowers which have given the French name of the plant (*mauve*) to that particular tint.

Its upright stems are well covered with stiff hairs, and its large leaves are cut into five or seven lobes, and toothed. The flowers are about an inch and a half across, and the arrangement of stamens and styles is as we have already described.

The Dwarf Mallow (*M. rotundifolia*) is similar, yet quite unlike. That is to say, any person acquainted with the Common Mallow, yet making no pretence to botanical knowledge, would say at once on seeing the smaller species, "That is a Mallow;" but the *downy* stems lie along the ground, the leaves have a rounder outline, and, like the much paler, inconspicuous flowers, are much smaller. There are slight differences in the essential organs correlated with these different sizes and habits, which are really very instructive when we come upon them in the species of so small a genus as this. In the Common Mallow the pyramidal cluster of anthers shed their abundant pollen before the stigmas are mature, and insects coming for honey alight upon this prominent central column and get well dusted with pollen. After the pollen is shed, the stamens curve downwards to be out of the way, the stigmas become mature, and separate somewhat, so that *they* form the alighting-place now; and should an insect come pollen-laden from a younger flower, some of its load is sure to adhere to the stigmas. In the Dwarf Mallow a slight variation produces different results. The anthers and stigmas mature at the same time, and the long styles curl over and intertwine among the anthers in such fashion that self-fertilisation is a certainty, almost before the flower opens.

Stamens of Common Mallow

Both these plants grow in similar situations, under like conditions as to soil, aspect, etc., so that we have here an admirable opportunity of testing the difference made by cross-fertilisation. Common Mallow aims at the good offices of flying insects such as bees, and its petals bear streaks pointing to the honey-glands at the base of the staminal column; its stems are covered with hairs to prevent crawling insects like ants approaching the flowers in an illegitimate manner, and foiling its schemes by plundering honey and pollen. The narrow base of the petals, too, is bearded to prevent uninvited insects getting illicit sips of honey, but the honey-bee contrives to get her tongue in and reach the honey after the flower has been closed for the day. Dwarf Mallow is not averse to such visits, and its hairs are reduced to a soft downiness, which would only interfere with slugs. But its possession of flowers so large and coloured points to the probability that it was once larger and cross-fertile only. It may, indeed, have originated as a variety of the larger species; and I am inclined to take this view. The smaller species is not so widely distributed in this country as the larger, but their range outside these islands is almost identical.

Stigmas of Common Mallow

Many species of bees, a few flies, and other insects regularly visit the Common Mallow, but in the case of the Round-leaved species very few insects — including the honey-bee — take notice of it.

The Tree Mallow (*Lavatera arborea*) during its first year of existence sets itself to building up a stout stem like a big cabbage-stalk, well furnished with large, lobed, roundish leaves. Next year it completes its growth, and has become five or six feet high, with flowering branches bearing glossy purple flowers an inch and a half across. This may only be seen very near the sea; indeed, the only places where it is truly wild are among the rocks and on islets off the coast. I have found it on island-rocks off the Cornish coast, where the only other vegetation was scurvy-grass, scentless may-weed, samphire, and a dock. There it raised its stout stem, and seemed to defy storm and sea alike.

GERANIUMS

THE Geraniums and Pelargoniums of our gardens and windows, so extensively cultivated among us, have made everybody familiar with the general form of flower in this family; but several of the native species are common, and from them we may get a truer notion of the relations of the organs. There is an Eastern legend to the effect that we are indebted to Mahomet for our Geraniums. Originally, the nearest approach to a Geranium was a Mallow. Mahomet had washed his shirt, and spread it out to dry upon some Mallows, and these blushed so intensely, owing to the honour done to them, that they became permanently crimson in hue, not merely the flowers, but to some extent the leaves also. The legend does not describe how the change in the structure of the flower was brought about; but if we compare a Geranium and a Mallow we shall find the greatest dissimilarity, however much superficial resemblance there may be—and that is really very slight. Many persons who are loth to believe that one species of plant may have been evolved from an earlier form by natural process, are quite willing to accept the foregoing legend as a reasonable explanation of the Geranium's origin.

We should like to take as our floral text the Wood Crane's-bill (*Geranium sylvaticum*), for no other reason than that it was this plant which more than a hundred years ago led Konrad Sprengel to undertake his researches into the relations subsisting between flowers and insects—researches that were neglected by both botanists and entomologists until Charles Darwin called attention to them, and supplemented them by his own observations. But the Wood Crane's-bill is confined to the northern half of our country, and therefore not readily accessible to the majority of my readers.

A more widely distributed plant, with larger flowers, is the Meadow Crane's-bill (*G. pratense*), which grows chiefly in damp meadows, but occasionally in woods also. It is a perennial plant, with upright branching stems, three or four feet high, swollen at the joints. Its leaves are five or six inches across, lobed and cut much like those of the Buttercup, on long stalks, and with slender stipules. Its striking flowers are an inch and a half in diameter, blue-purple in colour. These are really handsome flowers, and not less interesting. Each flower-stalk, or *peduncle* as the botanists term it, supports two flowers on shorter, more slender stalks, which are known as *pedicels*, to distinguish them from the others, but we may more familiarly speak of them as foot-stalks. These flower-stalks are covered with hairs pointing downwards, to prevent creeping insects reaching the flowers and stealing honey or pollen. There is reason for this precaution, and for the beard of hairs upon the lower part of every petal, to keep small flying insects from reaching the honey without rendering service. The five-celled ovary bears a long thick style divided at the top into five branches.

Now, when the flower opens, these five branches or stigmas are pressed closely together, with their stigmatic surfaces within. At this time the ten stamens lie flat upon the petals at right angles with the style; they are not quite mature. Then five of them erect themselves so that the anthers are close to the immature stigmas. They shed their pollen, and fall back to their original positions, whilst the other five erect themselves, shed their pollen, and also fall back. Now comes the turn of the stigmas: the five close-pressed arms of the style separate and stand widely apart, their stigmatic surfaces ready for the reception of pollen. The pollen of that flower is all gone, so that if the seed-eggs in the ovary are to be fertilised at all, it must be by pollen brought from another flower. Such a central support as is thus afforded first by the stamens, then by the stigmas, is at once taken advantage of by bees and butterflies as an alighting-stage whence they can extend their long tongues straight down to the honey-glands at the base of the ovary. I need not describe how the insect that alights on a young flower will get its under side well covered by pollen-grains, which will be detached by the adhesive stigmas when an older flower is visited. In all probability some visitors may prefer to alight on the petals, and these would, no doubt, obtain honey without earning it; though even such a loss may be prevented by the bearded claws making the honey difficult of access except from the central rostrum. Lubbock, following Sprengel, considers the office of these hairs on the petal-claws to be the protection of the honey from being weakened by rain; this is probably a correct view, but I am of opinion that their chief purpose is to act as a barrier to approach from the circumference instead of the centre.

I need not deal with the distinctions between the Meadow Crane's-bill and the other large-flowered species—the Bloody Crane's-bill (*G. sanguineum*) figured in the title-block to this chapter, with one of its stations, Kynance Cove; the Wood Crane's-bill (*G. sylvaticum*), and Mountain Crane's-bill (*G. perenne*) —although there are differences sufficient to warrant the systematist in giving them different names, the arrangements and movements of stamens and stigmas are very similar in each of these. They are all perennials, and they have all become so adapted to insect-fertilisation by the pollen of other flowers, that failing insect-visits they cannot produce seed. But there is another group of Crane's-bills that are annual or biennial in duration, and that have small, less-widely-spread flowers, and that have evidently receded from this absolute dependence upon cross-fertilisation.

The Mountain Crane's-bill may be regarded as a transitional species, showing how these others got their smaller flowers and their partial enfranchisement from insect-thraldom. In this case the stigmas ripen and expand before the second set of stamens have shed all their pollen, so that the chances of cross- and self-fertilisation are almost equal, for if insects do not immediately come along with foreign pollen, the anthers will place the pollen on the stigmas. The flowers of this species are only half an inch across! But in the Dove's-foot Geranium (*G. molle*), which grows abundantly among grass in pastures and waste places, the flowers are even smaller. The first set of stamens mature before the stigmas, but the latter spread themselves before the second set of anthers shed their pollen; so that self-fertilisation must often occur. The Small-flowered Geranium (*G. pusillum*) has flowers smaller still than the Dove's-foot; and here we find that the stigmas are mature before any of the stamens are ripe, and ready to receive it as soon as shed. Mark this point: self-fertilisation being a certain method of setting seeds with a few pollen-grains, only five of the stamens are complete—the outer five bear anthers, the inner five are merely filaments without anthers!

The Round-leaved Geranium (*G. rotundifolium*) is very similar to the Dove's-foot in size and appearance, but the petals are much paler, they are of narrower proportions, and they lack the "beard" on the claw: any insect can seek honey in the way that seems best, and push about among anthers and stigmas as it pleases, self-fertilisation being the end aimed at. This species is rare, and is only found in the southern parts of these islands. Herb-Robert (*G. robertianum*) is the most abundant of the genus, and its more or less reddened leaves and stems are conspicuous in hedge-banks almost throughout the year. The intensity of the redness suffusing the whole plant led our forefathers to dub it Robwort, that is Redwort, and no doubt by a process of elision familiar enough to those who have mixed much with our rural population, the *w* in Robwort was omitted, leaving Rob'ort to puzzle a later generation, that spelled it Robert, and wondered whether the name signified that the plant had been dedicated to a St. Robert. Linnæus regarded it as a personal name, and gave it a Latin termination by which the plant will continue to be known. The flower illustrates the ease with which an entire difference in the form may be effected. All the species of Geranium already described have *spreading* sepals— that is, they stand out at right angles with the foot- stalk, but in Herb-Robert and the Shining Crane's- bill (*G. lucidum*) the sepals maintain an erect position, so that they produce a false tube, as in Wallflower and some members of the Pink family. The stigmas and anthers mature together, but insects are still invited to help shake the pollen on the stigmas, or even to bring a little from another flower. We know they are still invited, because the petals bear honey-guides in the shape of streaks pointing down the tube, and there are no hairs upon the claws of the petals to interfere with their tongues reaching the five globules of honey.

So much for the flowers of the genus Geranium, which all share the habit of turning their faces to the sun: we shall still find a point or two of interest in the fruit and the mechanism for seed distribution. The ovary consists of five carpels which are ranged round a central axis growing up from the receptacle and between the styles. When the seeds and carpels are fully grown, the latter become separated from each other, and the five styles also separate from each other, though for a time they remain attached to the carpel (now open at the inner edge) and to the central axis; but by the continued lengthening of the last-named there comes a peeling off of the style, followed by its rapid curling up from the lower end and carrying the carpel with it. This act reverses the carpel completely, so that the contained seed is hurled out of the carpel and away. A few hairs at the bottom of the carpel-opening prevent the seed from falling out at the beginning of the curling movement.

Geranium scattering seeds

The fruit of Herb-Robert differs somewhat from that of the other species in the fact that the styles separate from the carpels, but a tongue-like process from the bottom of the style keeps it in place behind the carpel, and when all is perfectly ripe the sudden curl up of the style acts like the steel spring in a toy shot-gun when released by the trigger-pressure, and the carpel (which was merely attached to the axis by a few silky hairs) with its contained seed is shot off with great force. It is well worth while growing any of our Crane's-bills in pots in order to study this seed-shooting process.

Herb-Robert discharging its seed

These fruits before they begin to break up are supposed when turned about to resemble the head of a crane with its long bill—hence the popular name, and the scientific name derived from the Greek, *geranos*, a crane.

The Stork's-bills (*Erodium*), of which we have three native species, are very similar to the small-flowered section of the Crane's-bills—especially *G. pusillum*. They have been originally specialised for cross-fertilisation by insects, but they have partially given up their dependence on insects, with the result that their pink or rosy flowers are small, and five of their ten stamens never produce pollen. The Hemlock-leaved Stork's-bill (*E. cicutarium*) is the most frequent species, growing on gravelly wastes, and producing small umbels of rosy flowers, each less than half an inch across. The Musky Stork's-bill (*E. moschatum*) is of larger stature, but of very local occurrence; easily detected by passing the leaves through the fingers, when the strong odour of musk will tell you which species you have, whether the more purple flowers are there to be identified or not. Then in sundry places, near the sea especially, we may find the Sea-Stork's-bill (*E. maritimum*), which in point of rarity comes between the other two, but is much smaller, with minute leaves, and pale-pink flowers not more than one-eighth of an inch across—that is to say, the flowers are pink when the petals are present, but often these are missing.

The most interesting thing about the Stork's-bills is the clever bit of mechanism by means of which they bury their seeds. Where you find the Stork's-bills growing it is in considerable colonies, and this is due to the seeds being dropped at a slight distance only from the parent plant, and sown in the exact spot where they fall. In some respects the formation of the carpels and styles round a central axis is like that of Geranium, but the style retains its connection with the seed, and is lined on its under-side with silky hairs. The growth of the axis releases the seed from the carpel, and the style curls up in corkscrew fashion, which causes every hair on the back to stand out stiffly at right angles. So the pointed hairy seed is "hung at the yardarm," so to speak, the tip of the style resting on the tip of the axis, until a puff of wind releases it and carries all a foot or two away. Then a remarkable thing happens: naturally the seed-laden end of the hairy corkscrew reaches the ground first, and the pointed end sticks in the earth, whilst the hairs of the corkscrew stand out between blades of grass or stems of other lowly plants. Rain or dew becomes absorbed by the corkscrew, which consequently lengthens. But the lengthening drives the pointed seed more deeply into the ground, because the hairs of the style will not allow the other end to be extended. When, on the other hand, a drier atmosphere causes the corkscrew to contract again, the short hairs on the seed act as barbs and prevent its withdrawal from the ground, so the hairs on the style have to yield, which they will do in this direction. And so by alternate contractions and extensions of the style the Stork's-bill's seed is sown in the earth.

Stork's-bill seed

Another tribe of the Geranium family is represented in this country by the little Wood Sorrel (*Oxalis acetosella*), one of the most beautiful of our wild plants. We may find it in all its chaste beauty in April or May, creeping over the leaf-mould and the decaying tree-trunk in moist woods, especially on a sandy soil. The creeping rootstock is slender and knotted, of a pinkish hue, and the leaves start directly from it on long foot-stalks. They are of the conventional shamrock pattern, divided into three broad heart-shaped leaflets, each having a distinct fold along the middle, and coloured with purple on the under-side. These leaves possess a considerable amount of sensitiveness, even irritability. At night-time, or on the near approach of rain, each leaflet droops until the central "nerve" touches the leaf-stalk and the two halves of the leaflet are at right angles to each other. The object of this arrangement is doubtless that rain and moist air may get uninterrupted access to the rootstock and roots beneath. But if the plant be grown in a pot and shut up for an hour or so in a dark cupboard, the leaves will be found to have assumed the nocturnal position—that is, the leaflets will be all folded down to their foot-stalks. If now the plant is brought out and exposed to full sunlight, the leaves will at once assume the

Wood Sorrel

normal or daylight position. A slight concussion on the leaf-stalks will produce effects similar to darkness.

The flower of Wood Sorrel is very similar to that of Herb-Robert. The erect sepals produce a modified tubular effect in the lower part of the corolla, which is helped by the petals cohering above the "claws." The petals are white, delicately marked with purplish honey-guides. The five carpels are united to an axis, but the styles are free, and their tips only are stigmatic. The flower is designed for insect-fertilisation, but no doubt often does without, for the flowers are like the leaves in that they close up at night, and then the stigmas are almost certain to get pollen attached. But in addition to these showy flowers displayed in spring, the Wood Sorrel produces never-opening (*cleistogamous*) flowers like those of the Violet, already described, only in this case these are little twisted petals which remain unexpanded until the swelling of the seed-capsule throws them off like a little cup. Within these cleistogamous flowers only five of the stamens produce pollen, and these are attached by hairs to the summit of the ovary close to the exceedingly short styles.

No matter by which kind of flower the seed-vessel is formed, it is altogether different from those of the Crane's-bills and Stork's-bills. The carpels remain attached to the axis, and each contains two or three seeds of a flattened rugged character, and red in colour. Each seed is wrapped in a white envelope, which is brought up to the opening of the ripe carpel, and if the plant is touched, or the pot it is growing in is jarred, the seeds are ejected with great force from the elastic envelope and thrown to a distance of several yards. When growing the Wood Sorrel, or the yellow-flowered *Oxalis corniculata* (a naturalised plant), in the greenhouse, I have amused visitors by causing a fusillade from these vegetable sharpshooters and shown the conspicuous seeds on leaves of other plants all over the greenhouse. The leaves of Wood Sorrel contain a considerable amount of binoxalate of potash, which has formerly given them a vogue as salad ingredients and as medicine.

The Balsam (*Impatiens*), though closely related to the Crane's-bills and Wood Sorrels, has its petals and sepals quite differently arranged; in fact, whereas those already described were regular in their shapes, the Balsams are very irregular. Nominally we may say there are five sepals, but two of these are mostly absent, and when present are reduced to very minute proportions. One of the remaining sepals is greatly enlarged, funnel-shaped, and ending in a slender hollow spur. There are but three petals, but it is considered that originally there were five. As the flower hangs, one of these petals is above the opening to the funnel, the others one on each side. Now these side petals are each two-lobed, and there is every probability that the two lobes represent two petals which have become united at their base. The five stamens are united below and surround the pistil, which ends in a simple five-toothed stigma. These organs hang down a little way inside the flower, so that the bees which frequent it must brush their backs against them.

Our native species is a succulent, thick-jointed annual, known as the Yellow Balsam or Touch-me-not

(*I. noli-me-tangere*), with red-spotted flowers of pale-yellow hue. The stamens in this species mature before the pistil to favour cross-fertilisation, for which the position of the organs is admirably adapted, for a bee entering the roomy sepal chamber must dust its back with pollen, and on visiting a somewhat older flower must pollinate the stigma. Yet the arrangement has not attained to that high degree of specialisation which would absolutely prevent self-fertilisation, for though Mr. Darwin covered up some flowers with fine net to exclude insects, eleven of them produced seed-capsules with a few good seeds in each.

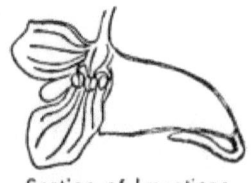
Section of Impatiens flower

The handsome Jewel-weed, or Snap-weed (*I. fulva*), that has made its way from North America and settled along many of our streams, has orange flowers, with the spur abruptly curved under the sepal sac. This is said to be so highly specialised for cross-fertilisation that even when artificially pollinated the pollen has no effect on the stigma of the flower that produced it. But to guard against the accidents to which such highly specialised flowers are subject, both these species produce minute *cleistogene* flowers, like those of the Violet and the Wood Sorrel.

Like the other branches of the Geranium family, Impatiens produces seed-capsules of the spring-gun order, and it is to these startling contrivances that the names Touch-me-not and Snap-weed are due. The seeds are attached to a central axis, and the five valves of the capsule are

Seed-vessel of Impatiens

attached to the two ends of this. As the fruit becomes ripe these valves become very elastic and tense; there is a strong pulling at the axis, but as the valves are all pulling in different directions they mutually neutralise such efforts. But they are very irritable, and a slight touch upon the capsule or its stalk will cause the five valves at once to make this pulling effort and sever their connection with the tip of the axis. Each coils up with a sudden snap, and the seeds are scattered far and wide.

Capsule valves coiling up and discharging seeds

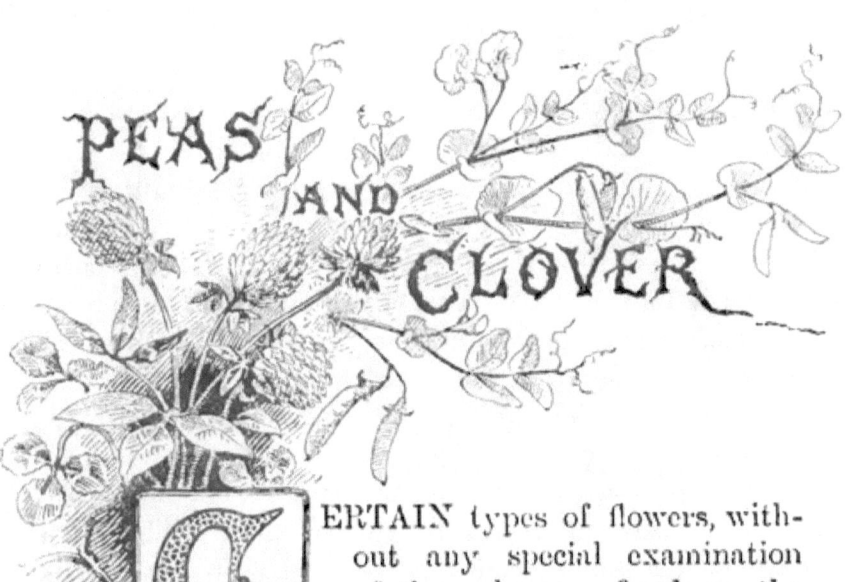

PEAS AND CLOVER

CERTAIN types of flowers, without any special examination of them, become fixed on the mind seven of the non-botanical; and that prevailing throughout the great Pea family, with its 4700 species, is one that is very well known by familiarity with the beautiful Sweet Peas of the garden, the Furze of the common and waste, and the Vetches of the hedgerow. It will not be possible within the limits of this chapter to pass in review the whole of the seventeen native genera comprised in this important family, but we may glance at the general structure of the flowers, the more important variations of it, and the principal kinds of flower-grouping (*inflorescence*).

First, however, there is a point in the history of these plants upon which I should like to touch briefly. The seeds of many of these leguminous plants (pulse) have for ages been known for their valuable feeding qualities, differing from grain in the fact that they contain a large percentage of nitrogen as contrasted with the predominating carbon compounds of the cereals. The question whence this great quantity of nitrogen was derived was long a problem to agricultural chemists, for it was found by the valuable experiments of Lawes and Gilbert at Rothamsted that whilst in most crops the nitrogen they yield can be accounted for by the amount of manure supplied to the land, with the addition of that contained in rain-water, when it came to an analysis of Peas, Beans, Vetches, etc., an excess of nitrogen was found beyond what could be accounted for in this manner. The only explanation that occurs is that the free nitrogen existing in the atmosphere is laid hold of and assimilated by the growing plant; but experimental botanists of high authority have been agreed that the plant has not this power, that though nitrogen enters freely into the air-spaces in the leaves the plant cannot avail itself of this element—it can only extract it by root-action from the soil in the form of ammonia and nitric acid.

One of the most startling of the recent disclosures concerning the micro-organisms was the discovery of bacteria capable of living in soil utterly devoid of all animal or vegetable matter, and by their feeding upon the purely mineral constituents of the soil, and passing them through their bodies to give such

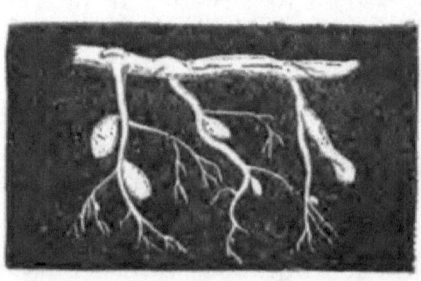
Bacteria nodules on roots of White Clover

a degree of fertility to the soil that green plants can sustain life in it. Following upon that important addition to our knowledge of the production of nitrates in the soil, came the discoveries of the actual species that do the work, one having the power of producing nitrous acid, and another capable of turning that nitrous acid into nitric acid. To these insignificant organisms is the fertility of our soils really due.

But what, you may ask, has all this to do with Peas and Clover? Everything: Peas and Clover could not provide the animal world with their valuable flesh-forming seeds but for these nitrifying bacteria—neither, indeed, could any of the ordinary farm crops be produced.

Most people who have grown Scarlet Runners, and have pulled up the roots after the frosts have cut down the stems, have noticed how these roots are all distorted and gouty in appearance. Similar nodules may be found on the roots of Sain-foin, Clovers, etc., and it is now found that these tubercles contain a great quantity of nitrogen and swarm with bacteria. The explanation, then, of leguminous plants containing more nitrogen than was in the soil appears to be that they take in the free nitrogen of the atmosphere and convey it to the roots, where the action of the bacteria converts it into such form (nitric acid) that the plant can assimilate it. It appears probable that each type or genus of leguminous plants has its own special bacteria.

Having thus briefly indicated a remarkable character of this group, let us look at the stem, leaves, and flowers. The most widely distributed of our native Peas is the Meadow Vetchling (*Lathyrus pratensis*), a very pretty plant with yellow flowers that may be found in hedge or copse in almost any part of the country. From the creeping rootstock several weak stems arise which have the appearance of being very leafy, but one-half of the apparent leaves are really enlarged stipules. The true leaves have been broken up into a pair of leaflets, and between these the leaf-stalk is continued some distance, and probably branches into several very slender tendrils which curl tightly round the stems and leaf-stalks of neighbouring plants. The original Lathyrus —like some existing species— most likely had several pairs of leaflets, but was under the necessity for devising means for hauling its weak stem up to the light, so the upper ones were reduced to the mere midrib which became a tendril— in some cases branched. In the Peas and Vetches all kinds of variations in the leaflets and the tendrils will be found, with enlargements of the stipules to make them serve as leaves where these have been converted into tendrils. In this species it will be noticed that the stipules are very large and conspicuous, and of the shape known as arrow-headed, due to their prolongation backwards

Meadow Vetchling

into a couple of slender ears. From the axils of the leaves a cluster of bright yellow flowers, varying in number from three to a dozen, are borne upon a long stalk, and each of these consists of a five-toothed tubular calyx and five petals. Now these petals are so different in form and size that they have received distinguishing names. The uppermost of the five is very much larger than the others, and is called the Standard (St). In the unopened flower it embraces and protects the smaller petals—the small size of the calyx-lobes making them useless for the purpose—and in the expanded blossom is the most conspicuous feature. Then there are two side petals known as Wings (W), and lastly two that occupy the central line of the flower and that are joined together by the lower edges; these are mentioned collectively as the Keel (K), because they have something of a boat-shape in their united condition. There are no less than ten stamens, and nine of these have the basal portions of their filaments united into a tube within which lies the pistil. The reason for the tenth stamen being free is that a gap shall be left in the upper side of the tube so that bees can get their tongues through to the honey which is secreted by the inner wall of the tube. On each side of the keel near its base there is a little

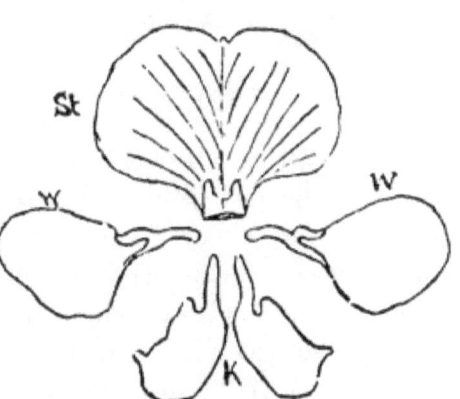

Parts of a Leguminous Flower

swelling corresponding with a depression in each of the wings, so that the whole interlock. The anthers shed their pollen before the stigma is mature, and it lies in the point of the keel.

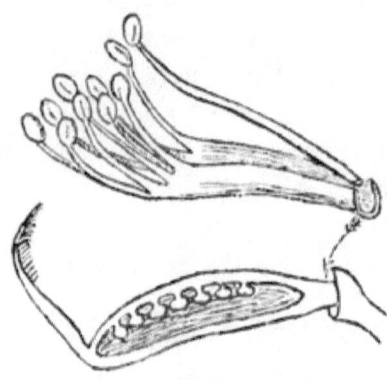

Stamens and Pistil of Pea-flower

When a bee alights upon the wings, its weight depresses them, and they bend the keel in such fashion that the curved pistil is pressed out at the tip of the keel, and the hairs with which its inner face is covered sweep out the pollen against the bee's abdomen. When the weight of the bee is removed, the organs resume their former positions; then if the bee visits a slightly older flower, the stigma will be pushed out against the bee's abdomen, and receive some of the pollen-grains from the first flower. Thus fertilised, the long flat ovary of one carpel develops into the one-celled *legume* which is characteristic of the family, and to which it owes its name. This is one of the best examples of the true nature of carpels one could have; for if some large species like the garden Pea be selected, and a nearly full-grown juicy legume be split open along *the back*, it will be seen to be just a folded leaf bearing seeds along its edges.

Pea-pod opened from back

The form of flower we have described at length is found with slight modifications to prevail throughout the whole family—that is, so far as the native genera are concerned. They are fertilised by hive and humble bees, though some of these are not able to get at the honey by fair means, and are driven to the expedient of biting a hole through the side, where they can get at it illicitly. There are some noteworthy variations in the matter of leaflets, some species having a great number, whilst in others they are reduced to tendrils or absent altogether. The Yellow Vetchling (*Lathyrus aphaca*), for instance, bears leaves only as a seedling, and these are divided into two oblong leaflets. Rarely, indeed, they may appear on an older plant, and then the leaflets are longer and narrower. But their absence is made up by the very great development of the stipules. The flowers are solitary.

The case of the Grass Vetch (*L. nissolia*) is worse; it has neither leaves, leaflets, nor tendrils. It is true it possesses leaf-stalks, and it has to make the most of these by flattening them out until they somewhat resemble the blades of grass, and use them as leaves.

Grass Vetch

The stems of the plant are very slender, and the crimson flowers solitary. No doubt when the species took to living amid grass by the sides of fields and similar situations it found that oval leaflets made it too conspicuous, so that it was eaten in preference to the grass, and also that they were not so well adapted for the struggle for existence as the leaves of the grasses. Then it got rid of its leaflets, and flattened out its leaf-stalks until they resemble a slender and fine-pointed grass-blade. Now, until it puts forth its flowers none but the botanist would suspect that it was anything but a grass. This species, it should be observed, is fully capable of self-fertilisation, and some of the flower-buds never expand. The Narrow-leaved Everlasting Pea (*L. sylvestris*) has two long and narrow leaflets, sickle-shaped stipules, and branching tendrils; throughout its length the stem has a flat expansion or wing on each side. This is a very near relation of the Broad-leaved Everlasting Pea (*L. latifolius*) of gardens, which is probably only a South European variety of it.

The ingenuity of the flower in adapting its structure so that it may profit by the visits of the bees is sometimes set at naught, or its practical value minimised, by the artfulness of the bee. Take the case of *L. sylvestris* as one of many examples that have been recorded. The situation of the honey within the staminal tube has already been indicated, and in the tube near the base two rounded orifices have been left for the passage of the bee's trunk. Mr. Francis Darwin found that sixteen out of twenty-four flowers of this species had the left passage larger than the right one. The humble-bees to save time bite a hole through the standard, and Mr. Francis Darwin found that *they always make this hole to correspond with the left passage*, because it is usually the larger of the two. He remarks: "It is difficult to say how the bees could have acquired this habit. Whether they discovered the inequality in the size of

Narrow-leaved Everlasting Pea.

the nectar-holes in sucking the flowers in the proper way, and then utilised this knowledge in determining where to gnaw the hole; or whether they found out the best situation by biting through the standard at various points and afterwards remembered its situation in visiting other flowers. But in either case they show a remarkable power of making use of what they have learnt by experience."

The general arrangements of *Lathyrus* are repeated in the true Vetches (*Vicia*); their habits, too, are much the same, most of the species growing in fields and hedges; though some, like the Bitter Vetch (*Vicia orobus*) and the Wood Vetch (*V. sylvatica*), frequent rocky woods. The Sain-foin (*Onobrychis sativa*), which is usually found only as an escape from cultivation, is an erect-growing perennial with large leaves, as much as half a foot in length, and divided into as many as a dozen pairs of leaflets. Its compact, erect racemes of rosy flowers are very ornamental, and prove attractive to bees not only for this reason but also because they are well provided with honey, the road to which is indicated by a number of crimson lines drawn down the standard. Although so striking in appearance, the flowers of Sain-foin are all defective, in the sense that the wings are in a minute undeveloped condition. The free stamen, too, is much shorter than the rest, and the staminal tube is covered above as well as below by the keel. When a bee alights upon the keel and presses it down, not only is the pollen pushed out but several of the stamens also. In this species the pod is curiously semicircular, downy, and netted, containing only one seed.

But this is not so curious as the pod of the Horseshoe Vetch (*Hippocrepis comosa*), which is flat but serpentine, with the upper margin notched. When ripe, this quaint pod breaks at the notches into from three to six horseshoe-shaped joints, each containing a single seed. The leaves are similar to those of Sain-foin, though with fewer leaflets; and the yellow flowers present a resemblance to the more plentiful Bird's-foot Trefoil (*Lotus corniculatus*), which gets its name from the fact that when three or four of the inch-long cylindrical pods radiate from the common stalk they resemble a bird's foot. But this species must not be confused with the rare Bird's-foot (*Ornithopus perpusillus*), whose jointed pods present a closer resemblance to the scale-clad toes of a bird.

The common species of Lotus already mentioned is a convenient, because readily accessible, example to take of those non-climbing leguminous plants that have large and distinct flowers. The leaves are here divided into five leaflets, though they appear to be trefoils, the lowest pair of leaflets being near the base of the leaf, and the stipules very minute. From three to ten flowers are borne at the end of a long flower-stalk, and individually these have such very short foot-stalks that the cluster forms a "head," or capitate cyme, to express the same thing technically. The anthers discharge their pollen into the keel before the petals have attained their full size, and by the time the bud expands the anthers have shrivelled up, but five of the filaments have continued to grow and thicken so that they press the pollen into the tip of the keel and keep it in position. Up through this mass pushes the lengthening pistil, and when a bee alights upon the wings and depresses them and the interlocked keel, the pollen is forced out on the bee's under-side. On a second or third visitor arriving, the now mature stigma is pressed against a similar part, in all probability bearing pollen brought from younger flowers. The pistil and pollen (if any left) fall back into their former position when the weight of the insect is removed.

The Kidney Vetch (*Anthyllis vulneraria*) has similar mechanism for fertilisation, yet there are two or three interesting differences in the flower-parts. The calyx is inflated and tubular, and so to avoid a waste of material the petals all have long claws, but of course such an arrangement in a low-growing plant leaves the honey exposed to such small creeping insects as may effect an entrance to the calyx. To guard against such a danger, the calyx is covered with woolly down, and even the leaves are covered with silky hairs to discourage them at the outset. Another result of the tubular shape of the flower is that only long-tongued bees can reach the honey.

The Rest-harrow (*Ononis spinosa*), though of similar structure, does not produce honey, therefore there is no free stamen, for there is no necessity for leaving an opening to the staminal tube. Yet the flower is fertilised by bees, and by bees only. It would appear that though they get no honey they have not yet learned to avoid so attractive a flower; there can be little doubt that in this case the rosy tint and large size of the flower lead them to expect more than they find. According to the observations of

Müller, it appears that the male bees on finding they have been deceived go off in disgust; but the more thrifty females, after assuring themselves there is no honey, start collecting pollen, and so get something for their trouble. There are two forms of this species—one, more or less prostrate, is covered with sticky and evil-smelling hairs; the other, erect, not sticky or fœtid, but armed with spines. Now this difference has evident relation to habit, the plant being rendered offensive in the one case to creeping things, and in the other to larger browsing animals. That the spines do make the erect form objectionable is no new discovery, for the ancient Greeks believed that no animal but an ass was foolish enough to attempt it, so they called it *Ononis*.

The Dyer's Greenweed (*Genista tinctoria*) is also without honey, and the ten stamens are all united, yet bees come to it for pollen, which is shed in the keel before the flower opens. As a fact, it may be said that the flower does not fully open until the bee alights upon it, for until then the standard is scarcely raised and the wings are locked to the keel, which in turn keeps the long curved pistil in a state of great tension. The weight of the bee causes both wings and keel to drop away, the stamens and pistil spring up to the standard, and the pollen falls in a shower over the bee. On the bee flying off, the wings and keel do not recover their former position, but hang down, whilst pistil and stamens remain close to the standard, so that on pollen-covered bees seeking more pollen they are likely to leave some on the stigma. Henslow found that if protected from insect-visits the flowers never open.

Rest-harrow.

The Furze (*Ulex europæus*) differs from Genista in possessing a very large yellow calyx, divided into two lobes and covered with dark hairs. The wings interlock with the keel, and the flower explodes like Genista when a bee alights, tempted by the sweet odour which the flower exhales. The Furze, of which we have also a smaller species, the Cat Whin (*U. nanus*), is interesting as being a large evergreen flowering shrub without leaves, but which is exceedingly prosperous in spite of that defect. As a young seedling it puts out a few of the trefoils with which its ancestors were doubtless clothed, but which later generations have been glad to exchange for sharp spines wherewith to prick the muzzles of browsing mammals that would otherwise have ended the race. The larger spines are similarly metamorphosed shoots, but many of the smaller ones still retain their three-parted character, for on either side a still smaller spine is produced. Another point of interest will be found in the little black hairy pods which succeed the flowers and contain the hard, shiny seeds. When these are perfectly ripe and the pods quite dry, the two valves separate with a snap, and each curls up so quickly that the seeds are flung away with force. On any sunny summer's day, if one is sitting beside a Furze-clump, the crackling of

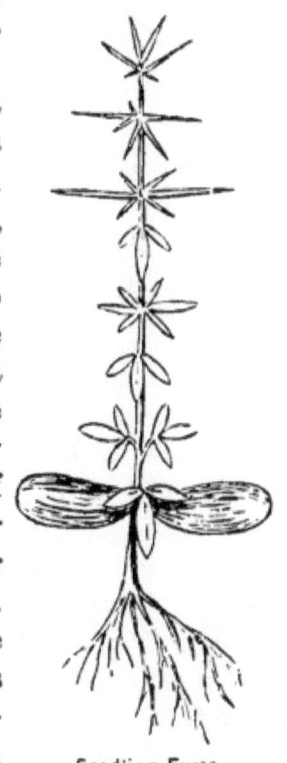

Seedling Furze

these bursting pods may be heard incessantly, and the fusillade of seeds distinctly felt.

It must not be presumed that all these plants producing pods (legumes) like the Furze exhibit similar elasticity of the valves; in some species the pod opens along one side only, in others not at all, but becomes detached from the stem and drops to the ground, where the decaying of the pod sets free the seeds.

The Broom (*Cytisus scoparius*) has flowers similar to those of Genista, but larger than those of Furze, and like both, honeyless. These honeyless flowers, as Mr. Darwin has remarked, are so constantly visited by bees that it is not reasonable to suppose they get nothing for their pains; it appears to me probable their tongues suck some pleasant moisture through the tissues of the staminal tube, just as in the Orchids they suck the walls of the spur. Without the visits of bees the Broom-flower never opens its keel; but bees are especially fond of it, and may be constantly seen about a bush when in flower. To quote Mr. Darwin: "When a bee alights on the wing-petals of a young flower, the keel is slightly opened, and the short stamens spring out, which rub their pollen against the abdomen of the bee. If a rather older flower is visited for the first time (or if the bee exerts

Broom

Broom, 1st condition

great force on a younger flower), the keel opens along its whole length, and the longer as well as the shorter stamens, together with the much elongated curved pistil, spring forth with violence. The flattened, spoon-like extremity of the pistil rests for a time on the back of the bee, and leaves on it the load of pollen with which it is charged. As soon as the bee flies away, the pistil instantly curls round, so that the stigmatic surface is now upturned, and occupies a position in which it would be rubbed against the abdomen of another bee visiting the same flower. Thus, when the pistil first escapes from the keel, the stigma is rubbed against the back of the bee, dusted with pollen from the longer stamens, either of the same or another flower; and afterwards against the lower surface of the bee, dusted with pollen from the shorter stamens, which is often shed a day or two before that from the longer stamens. By this mechanism cross-fertilisation is rendered almost inevitable." It will be noted that Broom is without a spine, and its leaves are of the three-foliolate character seen in the seedlings of Furze. Furze has no doubt been evolved from a Broom-like plant.

Broom, 2nd condition

In the Lucerne genus (*Medicago*) the flowers are honeyed, and therefore one of the stamens is free, to allow access to the honey. Each wing is united to the keel by a couple of processes, one pointing forwards and the other backwards, and locking into a depression in the keel. These points are most easily seen in the cultivated Lucerne, or Purple Medick

(*M. sativa*), because of its larger flowers, which can only be operated upon by humble-bees. The irritability of this species is located in the staminal tube, and on this being excited by the bee's tongue, the staminal tube and pistil are violently curled up to the standard, and the pollen showered over the bee, whilst the keel and wings become permanently depressed. Should no humble-bee arrive, the flower does not open, but self-fertilisation takes place. The pods of this genus are spirally coiled helix-fashion, but differ among themselves as to the number and closeness of the coils. Those of the Spotted Medick (*M. maculata*), which has been recommended as a fodder plant, are so sharp and so numerous that cattle often refuse to feed in the fields where they grow. The Melilots (*Melilotus*) have small honeyed drooping flowers in long racemes. Their general behaviour is much like that of Sain-foin.

So far as we have gone we have seen great differences in the flower-grouping of the Pea family, from the large distinct flowers of Broom, the loose racemes of Genista, the dense racemes of Sain-foin and Lucerne, the loose heads of Lotus, the denser heads of Kidney Vetch, until finally we reach the very compact heads of the Clovers (*Trifolium*). A novice who was confronted with a Clover-plant and a Pea both in full flower might well be pardoned for his failure to see a near relationship between the two. Yet if he is instructed to pull the head carefully to pieces and examine a single flower from the group, he will find in all essential points it is like that of the Pea, but the petals form a tube specially adapted to the long tongues of certain insects. Some of these flowers—

as the Red Clover (*T. pratense*) and the White Clover (*T. repens*)—appear each to depend upon some special insect for fertilisation.

Years ago, in *The Origin of Species*, Mr. Darwin told how he had experimented with bees and Clovers to ascertain how far the latter were dependent upon the former for fertilisation. He said: "Twenty heads of Dutch Clover (*T. repens*) yielded 2290 seeds, but twenty other heads protected from bees produced not one. Again, 100 heads of Red Clover (*T. pratense*) produced 2700 seeds, but the same number of protected heads produced not a single seed. Humble-bees alone visit Red Clover, as other bees cannot reach the nectar. It has been suggested that moths may fertilise the Clovers; but I doubt whether they could do so in the case of the Red Clover, from their weight not being sufficient to depress the wing-petals. Hence we may infer as highly probable that, if the whole genus of humble-bees became extinct or very rare in England, the . . . Red Clover would become very rare, or wholly disappear. The number of humble-bees in any district depends in a great measure on the number of field-mice, which destroy their combs and nests; and Col. Newman, who has long attended to the habits of humble-bees, believes that 'more than two-thirds of them are thus destroyed all over England.' Now the number of mice is largely dependent, as every-one knows, on the number of cats; and Col. Newman says, 'Near villages and small towns I have found the nests of humble-bees more numerous than elsewhere, which I attribute to the number of cats that destroy the mice.' Hence it is quite credible that the presence of a feline animal in large numbers in a district might determine, through the intervention first of mice and then of bees, the frequency of certain flowers in that district!"

The Dutch or White Clover is fertilised by hive-bees, but further experiments have shown that if they are excluded the plant is not absolutely infertile, though it sets very few seeds; this indicates that

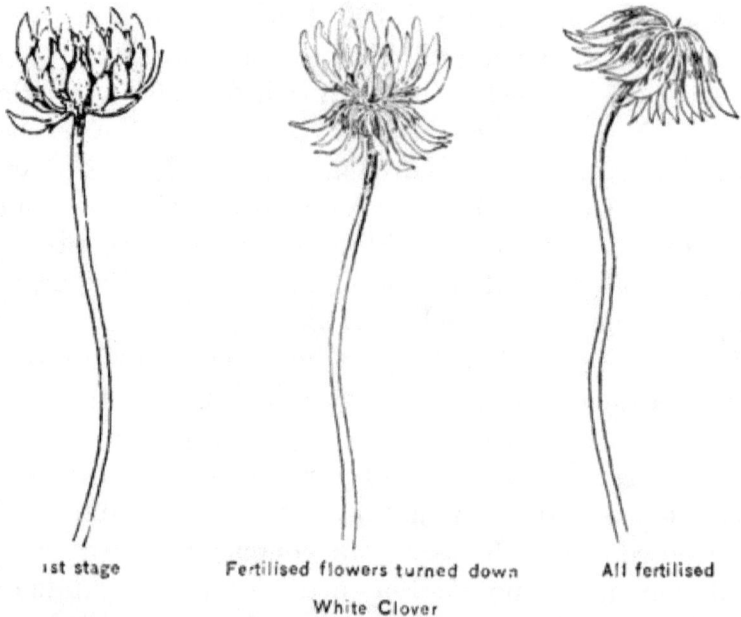

1st stage   Fertilised flowers turned down   All fertilised

White Clover

individuals exhibit a tendency to revert to self-fertilisation. The flowers are clustered in heads on a common stalk which is four or five inches high, but every separate flower has also its own little foot-stalk of which it makes important use. At first all the flowers stand erect, and the outer or lower ones of the cluster mature before the upper or inner flowers; then as the bees fertilise these they bend their foot-stalks, and the flower begins to dry up round the swelling ovary and to hang down round the common stalk. In this way the fertilised flowers never stand in the way of the virgin blossoms, and of these the mature specimens are always the outer row for the time being.

The three figures on p. 154 illustrate this well.

In the first of these figures one solitary flower to your left hand has been fertilised and has commenced to turn down; in the next about a dozen have so acted to be out of the way of the younger flowers; and in the third every one is fertilised and turned down. Here are also figures of individual flowers separated, of which this is a side view. At first the stamens are included in the keel, but when a bee alights upon this and the wings, pushing its head between the keel and the standard in order to reach the honey, the keel is so depressed that the anthers burst out and cover the bee's under-side with pollen.

White Clover
Side view

When the bee departs and the pressure is removed, the stamens retire to the keel again, and repeat the process with other bees until the pollen is exhausted. Then the pistil lengthens and the stamens dwindle, so that the stigma comes against the bee on the next visit and gets fertilised. The top figure on the next page shows the upright flower before the bee depresses the keel, and the lower one shows a flower from which the calyx and the standard have been removed to give the upper view of the flower with the entrance to the honey on each side of the style.

The Red Clover appears to be fertilised almost exclusively by humble-bees in this country, but Hermann Müller says he has often seen the hive-bee in Germany forcibly breaking open the flowers for the sake of the honey and pollen it could not obtain legitimately; and in New Zealand, in the absence of humble-bees, the plant appears to have adapted itself to self-fertilisation.

Among those native species that are not cultivated by the farmer is the Subterranean Trefoil (*T. subterraneum*), a hairy plant, with cream-coloured heads containing only from one to three perfect flowers surrounded by imperfect calyces, each of which has five rigid lobes. When the flowers have become fertilised and the pods begin to grow, the flower-stalk lengthens until the head touches the soil. Then the aborted calyces develop into long fibres

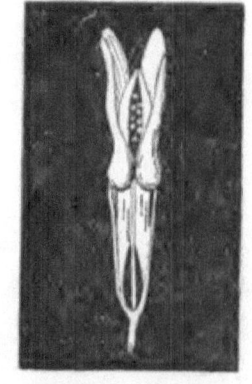

which enter the soil, and the lobes at their tips separate and spread out in such manner that the pods are pulled down into the ground. The hairs of the calyx absorb moisture from the earth and feed the ripening seeds — which will not ripen if this burying process is prevented. In this species the flower falls off as the pods begin to enlarge, but in

Hare's-foot Trefoil.

most of the Clovers it remains round the seed-vessel till this is ripe.

The roots of the Clovers bear many nodules produced by the nitrifying organisms, so that a field that has been cropped with Clover is rich in nitrogen, and farmers find it is particularly advantageous to follow Clover with a crop of wheat.

SUNDEWS

SUNDEWS form one of the smallest of plant families so far as our native species are concerned; but it is not on that account one of the least interesting. The Sundews grow naturally in wet places on moors; especially where there is a bed of sphagnum-moss growing on a gentle slope with water percolating slowly through it. There, with their simple roots among the sphagnum, deriving probably nothing but water from it, they occur in great numbers, though scarcely noticeable by the average rambler. The commonest form is the Round-leaved Sundew (*Drosera rotundifolia*), which is widely distributed throughout our land, and therefore to be obtained without great difficulty by most of my readers, if they will look for it in the places indicated. The leaves radiate from the rootstock and lie almost flat upon the ground, or the surface of the sphagnum. The blade of the leaf is circular, and the leaf-stalk is equal in length to about three diameters of the blade.

The remarkable feature of the plant consists of a large number of appendages fringing these leaf-blades and studding the upper surface. These are often spoken of as hairs, but though they may have a similar origin to hairs, they are too fleshy to make such a description of them anything but misleading. They are of a crimson tint, and end in enlarged glandular tips which excrete a clear, sticky fluid like liquid gum. When the leaf is fully expanded, all these tentacles, as they may be more fitly termed, bend away from the centre of the leaf, and in the sunshine, when each gland is surrounded by a globule of fluid, the leaves are no doubt mistaken for dewy flowers by small insects which alight upon them in the hope of finding honey or to sip the supposed dew—flying insects being thirsty creatures. But there is neither honey nor dew to refresh the insect; only its own death and dissolution to fatten the Sundew. The fly touches the tentacle-glands with its limbs, and is held by the gummy fluid. It seeks to free itself, but only succeeds in further irritating the glands, which pour out more mucilage and convey the irritation to the leaf generally. Thereupon, all the tentacles begin to bend over to the poor fly; their sticky tips adhere to its wings, its back, its sides, whilst the edges of the leaf turn up, and so convert the centre into a shallow basin, in which lies the fly held securely by a network of tentacles. Now there is poured out a digestive fluid which actually dissolves the soft parts of the fly, and when the process is complete the leaf absorbs this extract of fly to the manifest advantage of the plant.

Round-leaved Sundew leaf

The question naturally arises, Why should a plant develop a trait so characteristic of animal life, and so thoroughly opposed to the general notions of plant-life? The answer is to be found in the habitat of the plant. We have seen, in considering the Pea-tribe, that nitrogen is so liable to be washed out of the soil that we should scarcely expect to find much of it where a spring or the drainage from the moors above comes down unceasingly. In the latter case there would probably be a small percentage of nitrogen washed down, but the sphagnum would be most likely to secure it, and the Sundews have therefore given up competing for it through their roots, and instead have so modified their leaves that these organs are able to obtain it for them. It is a noteworthy fact that, though this carnivorous habit has been developed in plants belonging to families unrelated to each other, they all agree in the fact that they grow in bogs and ditches, or places where their roots are similarly steeped in water; and it is reasonable to suppose that this similarity of habitat has led the plants to adopt like means to obtain the necessary nutrition. Such a power could not be so well exercised if the specialisation of the leaf-form were not correlated with a remarkable degree of sensitiveness. This exquisite sense is restricted to the glandular tips of the tentacles, and it is not until some portion of the insect has come into actual contact with this that the neighbouring tentacles know anything of the matter. But no sooner does the leg of a fly press through the globule of mucilage and touch the gland itself than the excitement is communicated through the tentacle to the leaf substance, and so to all the other tentacles. It is not merely irritation that is thus communicated by one tentacle to another, but the knowledge of the direction whence the irritation proceeded. This is shown by the fact that the tentacles so apprised bend over towards the centre of disturbance.

It might be supposed that in so sensitive an instrument as the Sundew leaf must be, with its many tentacles, there would be a considerable amount of wasted effort owing to irritation being accidentally set up by grains of sand, drops of rain, etc., falling on the leaves; but this is not so. It is remarkable that so fine a sense is possessed by the glands that they refuse to take notice of any inorganic solids. Place a gravel-fragment upon the leaf, and the tentacles will take little notice of its presence; but if the foreign substance be the merest fragment of hair, not only will the base of the tentacles be inflexed towards it, but the fluid exuding from the gland becomes acid and of a digestive quality. This is accompanied by changes in the substance of the gland of the tentacle itself—the purplish colouring matter being gathered up into little masses suspended in a colourless fluid. When the excitement has ceased and the tentacles have resumed their normal position, it is found that this colouring matter has been again dissolved, and now gives its colour to the whole of the internal fluid. In Mr. Darwin's famous experiments on this plant he found a degree of sensitiveness greater probably than that possessed by any nerve in the human body, even when inflamed. A tiny morsel of cotton which weighed no more than $\frac{1}{5000}$th part of a grain was sufficient vegetable matter to cause the tentacles to bend over. It is not easy to mentally estimate the smallness of that cotton particle, but what idea can be gained of the size of a snippet of hair which weighed only $\frac{1}{78,000}$th part of a grain! Yet this inconceivably small portion of animal matter was equally effective in setting up the necessary irritation.

Such inorganic substances as are needed by plants as food, and such as they are capable of assimilating, do effect inflection of the tentacles and the aggregation of the protoplasm within. Mr. Darwin experimented with the salts of ammonia, and found them to have a powerful effect on the glands, causing inflection of the tentacles and aggregation of the protoplasm. The figures used above for cotton and hair appear huge, wholesale quantities when compared with some of the figures denoting the amounts of ammonia found sufficient to cause similar demonstrations of sensibility in the tentacles. Of carbonate of ammonia $\frac{1}{268,800}$th of a grain was needed, but with nitrate of ammonia $\frac{1}{651,200}$th was sufficient, whilst of the phosphate of ammonia the moderate amount of $\frac{1}{19,760,000}$th was ample!

These figures have only an indirect connection with the natural history of the plant, and are only mentioned to show the marvellous degree of sensitiveness possessed by the leaves, which have a curiously close resemblance in form and action to the Sea Anemones of our coasts. In nature the Sundew is concerned in catching and digesting small flies and beetles, and it is almost an impossibility to find a plant that does not bear proof of this fact in the majority of its leaves. Here the leaf is newly closed upon the victim, there the tentacles have returned to their widespread attitude, and the indigestible horny portions of its last meal are still to be seen in the centre of the leaf.

There is little of an interesting character known concerning the flowers of the Sundew, which are small, the white petals being but little longer than the sepals, and both stamens and pistil maturing at the same time. In many cases these flowers do not open, or open very slightly, but the pollen-grains send out shoots which reach the stigmas and so fertilise the seed-eggs.

The leaves of the three native species present variations in form and habit that probably refer to some differences in their victims. The round leaves of *Drosera rotundifolia* are extended at right angles with the rootstock; the spoon-shaped leaves of *D. intermedia* hold themselves erectly, and the longer, more slender ones of *D. anglica* are borne almost erectly. There is a striking parallel in two species of Plantain: *Plantago media* has broad, more or less rounded leaves, spread in rosette fashion on the soil, whilst *P. maritima* has long, slender, lance-shaped leaves which are borne nearly erectly. Sir John Lubbock considers that the attitude is determined by the shape of the leaves, and this may be so; but no doubt in the case of Drosera the difference would enable those with long leaves to catch more low-flying insects, whilst the flat, round leaves would be more likely to capture those that crawled over them.

PARSLEY AND CARROT

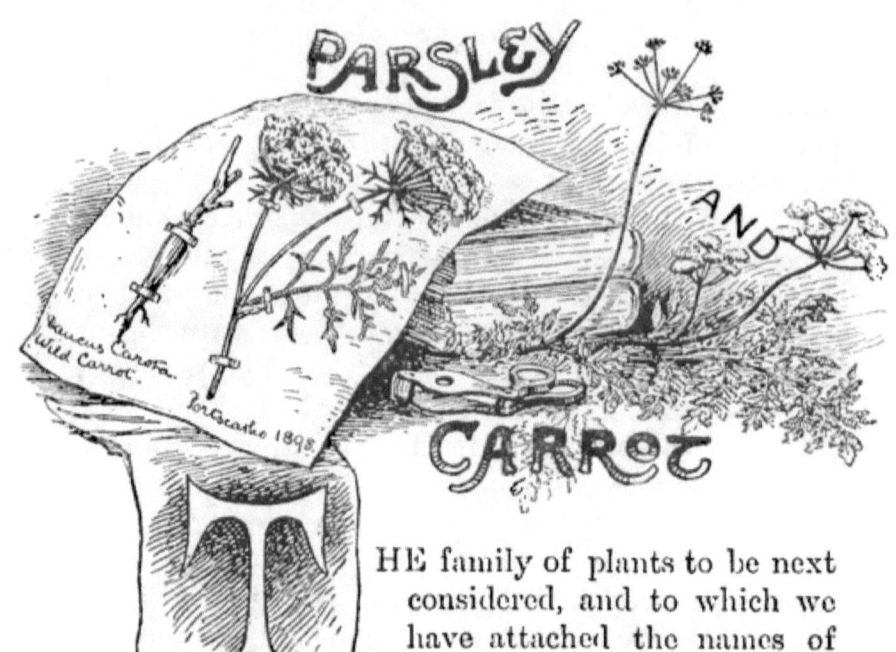

THE family of plants to be next considered, and to which we have attached the names of two well-known cultivated species, is a very extensive one, but also one whose members agree very closely in the appearance as well as in the structure of their flowers. Some groups already described have their flowers so specialised and adapted for fertilisation by a few particular insects that we may regard them as aristocratic and exclusive in their relations with the insect world; but the present family is thoroughly democratic—its flowers are formed and massed on popular lines. All insects, whether large or small, long-tongued or short-tongued, are made welcome. There is no need for special weight to depress certain petals, no need for strength to push open spring doors, no need for tongues constructed like elephants' trunks to reach down long and slender tubes, and

Cow Parsnip.

there are no impassable barriers of stiff hairs to keep off those that have the misfortune to be naturally small.

The flowers are always small individually, usually white, but so associated in flat heads that they are very conspicuous. In most cases, too, they produce honey, and the pollen is shed before the stigmas are ripe to receive it—showing that cross-fertilisation is desired. In consequence of this regularity of the flowers throughout the family, it will not be necessary to pay much attention to them after describing their general characters.

One of the best — because the largest — of the native species, for our purpose, is the well-known and plentiful Cow Parsnip (*Heracleum sphondylium*). A vertical section must be made through the flower, and then we shall see that its lower portion consists of a two-celled ovary, closely invested by the calyx, at whose upper margin we may be able to distinguish five minute teeth, but these are often absent. There are five white or pinkish heart-shaped petals, unequal in size. On top of the ovary there is a two-lobed disk on which the nectar is poured out, and to this are attached the two curved styles. There are five stamens, whose filaments curl over towards the centre of the flower, and from their tips are loosely hung the swinging anthers.

Each of these little flowers is mounted on a foot-stalk, and a variable number of these foot-stalks spring from the summit of a longer and stouter stalk. This flower-cluster is called an umbel, and if you look at it from below you will see that it bears considerable likeness to a Japanese umbrella. This peculiarity has earned for the family the name of Umbelliferæ, or the Umbel-bearers. Following the stalks of these umbels downwards, we find that a large number of them spring from a yet stouter, leafy stem, and as the flower-cluster thus consists of an umbel of umbels, it is described as a compound umbel. The majority of the plants in this family, as we know them in this country at least, have very beautiful leaves of the compound order: they are broken up into leaflets or lobes, and these again divided and deeply toothed, so that popularly they may be said to resemble fern-fronds. Most of them, however, are characterised by a soft, thin texture, which readily parts with its moisture and becomes flaccid. But we have still left to us species with much simpler leaves, and they probably exhibit to us some of the stages in the evolution of the highly compound leaf.

After the flowers of Cow Parsnip have faded, each of the foot-stalks bears a couple of little shields, each poised on a slender bristle attached to its summit. These are the fruits, corresponding with the two cells of the ovary, and each contains a single flat seed. These fruits are very characteristic of each species; in fact, they constitute the most important character for distinguishing between forms that resemble each other in foliage and flowers. They have normally five primary and four secondary ridges on their outer face, and between some of these ridges there are tubes pierced in the carpel-wall, and filled with the essential oil that gives a distinctive odour to the fruits of Anise, Carraway, Coriander, Cummin, Dill,

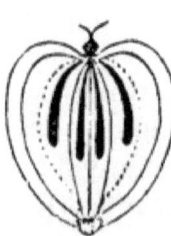

Fruit of Cow Parsnip

Sea Holly.

and others. The fruits of Cow Parsnip have four of these oil-tubes, or *vittæ*, as they are termed, on the outer face, and two on the inner. The fruits in this family do not split open to release the seed, as in most other cases, so that the well-known flavouring "carraway-seeds" are not merely seeds, but seeds plus seed-vessels.

If we visit almost any marshy place, we shall find a little plant, with roundish leaves, the stalk of which is attached to the centre of the under-side. This is commonly known as Marsh Penny-wort (*Hydrocotyle vulgaris*), but farmers and shepherds call it White-rot, in the belief that their flocks and herds eating it become subject to disease; the disease, however, is due to an animal parasite, the Liver-fluke, which abounds in marshy ground. The plant has a creeping white stem, which roots at intervals, and there sends up several leaves, and between them its simple little umbels of minute greenish flowers. Simple leaves of this kind are rare among the umbel-bearers, but on chalky ground in the south-east of our island grows another such, known as the Hare's-ear (*Bupleurum rotundifolium*). This has egg-shaped leaves, the broad end of which completely encircles the stem. In the language of the textbooks, the leaf is *amplexicaul* and the stem *perfoliate*. A novice would never take this plant to be an umbel-bearer, for the little umbels of a few tiny yellow flowers are each enclosed in from three to five large leafy bracts, which look like green petals, within which the actual flowers may pass as stamens. A closer scrutiny, however, will soon put one right on this point.

The well-known Sea-holly (*Eryngium mari-*

timum)—that grows with its roots deeply buried on sandy shores, so dry and hot that it seems impossible any plant could live in them—is a yet more puzzling plant to put into the hands of a budding botanist with a request that he should name its family at sight. The pale-blue flowers are gathered into dense heads with stiff bluish-green bracts interspersed among them, which give them a thistle-like appearance. This condition is brought about by the suppression of the usual foot-stalks to the flowers. Then, too, we get in this species an advance upon the simple round or oval leaf: the radical ones are here almost round, but the upper leaves are deeply lobed, and the lobes are arranged like the fingers on a hand (*palmate*); the margin of the lobes is thickened, and runs out into long spines like those of the Holly, and the plant is even more prickly than *Ilex aquifolium*.

The little Sanicle (*Sanicula europaea*), that grows in thickets and on the borders of woods, is another of these umbel-bearers that appear at first sight to be something altogether different. The umbels of pinkish flowers are very small, half-round, and the outer flowers contain stamens only. The leaves are an advance in complexity upon those already described: they are palmate, but each principal lobe bears smaller lobes, and the margins are regularly toothed. These leaves may now be compared with those of the Cow Parsnip, which are often as much as three feet long; they are distinctly broken up into leaflets, which are more or less pinnately divided, and the lobes toothed. By considering a series of leaves produced by different species in one family, we may get indica-

Wild Carrot.

tions as to the stages by which such beautifully complex leaves as Chervil, Parsley, Carrot, etc., have been developed.

Many of the family have interesting associations as the originals whence some of our table vegetables have been evolved by the market gardener; others have been long famed for their medicinal and aromatic qualities. Hemlock, the State poison of Athens, by which Socrates is believed to have met his death, is produced by our native *Conium maculatum*. Other highly poisonous natives in this family are the Wild Celery (*Apium graveolens*), Water Hemlock (*Cicuta virosa*), Water Dropwort (*Œnanthe crocata*). Cultivated Celery is earthed up not merely to increase the size of the leaf-stalks and make them white, but because the absence of light prevents the development of the poisonous principle.

Although it has been stated that Parsley (*Carum petroselinum*) is a native of Sardinia, whence it was introduced in 1548, Sir J. D. Hooker is of opinion that its native place is unknown, and that wherever it is now found growing wild it is merely an escape from cultivation. The Parsnip (*Peucedanum sativum*), on the contrary, is a true native of England, and has been cultivated from Roman times. In the wild plant the root is small and hard, but by cultivation it has been enormously increased in size, and with the enlargement has come a more succulent texture. Much the same may be said respecting the Carrot (*Daucus carota*). The wild plant produces a large rosette of really beautiful leaves from a rootstock that is little suggestive of edible qualities; we must suppose that when its virtues were first discovered

Wild Carrot.

tions as to the stages by which such beautifully complex leaves as Chervil, Parsley, Carrot, etc., have been developed.

Many of the family have interesting associations as the originals whence some of our table vegetables have been evolved by the market gardener; others have been long famed for their medicinal and aromatic qualities. Hemlock, the State poison of Athens, by which Socrates is believed to have met his death, is produced by our native *Conium maculatum*. Other highly poisonous natives in this family are the Wild Celery (*Apium graveolens*), Water Hemlock (*Cicuta virosa*), Water Dropwort (*Œnanthe crocata*). Cultivated Celery is earthed up not merely to increase the size of the leaf-stalks and make them white, but because the absence of light prevents the development of the poisonous principle.

Although it has been stated that Parsley (*Carum petroselinum*) is a native of Sardinia, whence it was introduced in 1548, Sir J. D. Hooker is of opinion that its native place is unknown, and that wherever it is now found growing wild it is merely an escape from cultivation. The Parsnip (*Peucedanum sativum*), on the contrary, is a true native of England, and has been cultivated from Roman times. In the wild plant the root is small and hard, but by cultivation it has been enormously increased in size, and with the enlargement has come a more succulent texture. Much the same may be said respecting the Carrot (*Daucus carota*). The wild plant produces a large rosette of really beautiful leaves from a rootstock that is little suggestive of edible qualities; we must suppose that when its virtues were first discovered our early forefathers were in a very low stage of civilisation—living chiefly upon roots, herbs, and wild fruits—or they would never have experimented with it. Yet the changed conditions with which cultivation has surrounded it, coupled with the careful selection of the most promising variations to serve as seed-bearers, have combined to give us what is practically a new plant. Few who are not intimately acquainted with the comparatively plastic character of most forms of life, and know how they may be modified by selection and changed environment, would imagine that the thin, dry stick of a rootstock found in the wild plant was the raw material from which the art of cultivation has produced such varieties as James' Intermediate and Long Horn.

Some species with large and striking umbels are visited by great numbers of insects by which cross-fertilisation is effected simply by walking over many of the flowers and getting well dusted with pollen, which is afterwards conveyed to other umbels. As we should expect to find from a consideration of the open disks upon which the honey is spread, this family is not favoured by long-tongued insects like butterflies, moths, and bees, yet swarms with flies and small beetles.

Woodruff and Goosegrass

CREEPING lowlily among the grasses of the pasture, its slender square stems spreading from the root, there may be found from spring to autumn the little Field Madder (*Sherardia arvensis*), with its terminal umbel of minute pink flowers. At regular intervals along the stem there are four or six lance-shaped leaves with bristly edges, and arranged in a whorl. On some of the stems we shall find that the lowermost leaves are in pairs only, and it is believed that the founder of this family was an opposite-leaved plant. A very large number of species like Coffea and Cinchona—none of them represented in this country, however—always have their leaves in pairs; and the contention is that what appears to be six leaves in a whorl on this stem of Field Madder is really only a single pair with a pair of leafy stipules from the base of each. The flowers are only about one-eighth of an inch across, and very like Jasmine blossoms in miniature. There is a calyx-tube with from four to six fringed teeth; a funnel-shaped tubular corolla with the limb cleft into four parts; four stamens and a two-armed style which protrude from the mouth of the corolla. Honey is produced on a disk at the bottom of the tube, and the stamens mature before the stigmas. When the pollen is shed and the anthers shrivelled, one arm of the style lengthens considerably, and the stigmas become ready for pollination, which must be effected by the visits of insects.

Woodruff (*Asperula odorata*) presents a similar appearance, but its stems grow erectly in woods and copses. Its firm leaves and stems are shiny, and as they dry after gathering, they give out a sweet odour of fresh hay. The flowers are as large again as those of Field Madder, more bell-shaped, and the calyx has no teeth; they produce honey, but the only result of insect-visits appears to be the shaking down of pollen from the anthers at the mouth of the corolla upon the two-branched stigma more than half-way down. Woodruff is insect-fertilised with its own pollen. When its two-seeded fruit is ripe, the calyx covered

Woodruff

with hooked bristles still invests it, and these little hooks catching hold of the fur of mammals or the feathers of birds secure the dispersal of the seeds.

There is a tendency in all these plants to develop flinty hairs or bristles on stems, leaves, or calyx; and in some the tendency is made to serve important ends. The true Madder (*Rubia peregrina*), from which the dye of the same name is prepared, has these in the shape of hooked prickles fringing the evergreen leaves, along the midribs, and the angles of the stems; but it does not develop them upon its fruit, which is a pulpy black berry much sought after by birds, as all black pulpy fruits are. The birds effectually disperse the seeds. The flowers in this species are greenish or yellowish.

The remaining native genus of these Madderworts consists of the species of Bedstraw (*Galium*), which exhibit the general characters of the Madders and Woodruff, the leaves being arranged in whorls, but the stems are much longer. Two of them—Lady's Bedstraw (*G. verum*) and the Crosswort (*G. cruciata*)—have yellow flowers to render them more attractive to beetles; the others have white or greenish flowers. There is no tube to the corolla, and the limb of the calyx forms a simple ring without teeth. The honey is secreted on a broad disk surrounding the branched style, whilst the four stamens are inserted between the lobes of the corolla.

The Crosswort may be readily known from the other yellow-flowered species by its leaves, which are only four in a whorl arranged crosswise. The Hedge Bedstraw (*G. mollugo*) is a soft, smooth species, with long stems that hang over hedge-banks and produce considerable masses of white flowers. There are rough irregularities of the stem-angles and the margins of the leaves. In the Field Bedstraw (*G. tricorne*) these roughnesses are developed into hooked prickles, which enable the plant to climb up the stems of the corn, and its roughish (but not hook-covered) fruits, which ripen about the same time as the corn, get efficiently distributed with it, so that the aid of bird and beast is unnecessary.

Goosegrass, or Cleavers (*G. aparine*), has reached a higher stage of development as a climber: its stems and leaves are extremely well furnished with the flinty hooks which enable its five-feet lengths of stem to scramble up and form green curtains over the hedges. But as this species cannot well enlist man's aid in dispersing its seeds, as the Field Bedstraw does, it covers its fruits also with the flinty hooks which catch readily and surely in fur or feather, and in any human garments. Scarcely a bird can alight upon that part of the hedge without taking a few away; and no mammal, whether mouse, rat, rabbit, stoat, fox, sheep, ox, or horse, can come to the hedge without picking up a number which will only be dislodged upon their pushing through or rubbing against another hedge.

Scentless Mayweed.

DAISIES AND THISTLES.

THE most numerous as regards both species and individuals, and the most widely distributed over the earth's surface, are the plants with flowers like those of the Daisy—members of the Composite family, as it is called. Many families of plants are chiefly found in some particular region: thus, of one family it will be said "Temperate and tropical regions, but chiefly of the Old World"; of another, "Northern temperate and arctic regions"; of another, "All temperate regions," and so forth; but the Composites—the Daisy family—are found in all regions. The family is split up into as many as 768 smaller groups (*genera*), which contain no less than 10,000 known species. The name of the family is founded upon the fact that what is usually termed the flower of any one of its members is really a densely-packed assemblage of flowers, so well compacted that they appear like a single blossom. As a matter of convenience such an *inflorescence* may be mentioned as a Composite flower.

In order that we may clearly understand the nature of these Composite flowers, let us revert for a moment to the Field Madder (*Sherardia*), and suppose that the little foot-stalks of its separate flowers were all suppressed, so that the flowers packed closely together stood on the expanded top of the stem with a whorl of leaves immediately below them. That would give an idea of the way in which Composite flowers probably originated. It would seem that there has been a tendency among small flowers of various kinds to get an advantage out of association. In glancing at the umbel-bearers we beheld one direction in which this tendency moved; there the tiny, and individually obscure, flowers by a loose kind of union became conspicuous, so that they attracted many small insects who fertilised great numbers of seed-eggs easily, and in return got abundance of pollen or nectar with little labour.

In the Guelder Rose (*Viburnum opulus*) the co-operative idea is carried further: the outer row of flowers in the cluster being developed to three times the size of the others, *but containing neither pistil nor stamens*, the available material being used up to the full for the purpose of advertising the flower-cluster as a whole. Here, then, we may find an example in plant-life of that altruism which Professor Henry Drummond believed to be the impelling law in the ascending evolution of the animal kingdom: a number of flowers sacrifice their natural function in order that they may thereby aid their sister-flowers to successfully complete their mission. This, as we shall see, is not the only example of altruism among flowering plants. Guelder Rose is not composite-flowered nor a true

Honeysuckle.

umbel-bearer, but an ally of Elder and Honeysuckle. We cannot stay to deal with the family, but we may glance for a moment at Honeysuckle (*Lonicera periclymenum*), because its cluster of flowers also shows a strong inclination to become a Composite. The calyces with their contained ovaries are crowded together at the end of a branch, and on examination it will be found that some of the calyces are joined together in pairs. The corolla has been drawn out into a long trumpet-shape, but opens unequally, not with the five lobes that denote the conjoined petals separate, but the limb of the corolla rather divided into two lips, one with three or four lobes, the other with one or two. The process of close association has begun, but at present only with the calyces and ovaries; the corollas stand as far as possible apart, and are usually half full of honey. The stamens and pistil, which extend beyond the mouth of the corolla, stand well away from each other. Only the larger moths and the long-tongued bees can reach the honey, and after the first drink only the moths can reach down far enough; but short-tongued bees come for the pollen which they can collect from the anthers easily, and in so doing no doubt effect cross-fertilisation, which is otherwise done by the fluffy faces of the moths as they push between anthers and stigma.

Now we will turn to the true Composites, and first glance at the simplest native form, the Hemp Agrimony (*Eupatorium cannabinum*), a plant that grows to a height of four or five feet in damp coppice and moist hedgerow. Its large handsome leaves are somewhat similar in appearance to those of the Hemp

(*Cannabis*), to which circumstance it owes part of its names. The pale-purple flowers are massed in heads that resemble the spreading clusters of the umbel-bearers, but if we gather one of these masses we shall find it is built up on an entirely different plan.

Taking off one of the main flower-stalks above the highest of the compound leaves, we find that at its next joint there is a pair of leafy bracts above which are several branches, each with a smaller pair of bracts, from which again there are five or six branches, but this time they end in a whorl of about ten overlapping bracts, forming an involucre to four or five tubular flowers which stand on a common flat receptacle.

Hemp Agrimony

Now, each of these ultimate clusters form what is known as a flower-head or composite flower, and the individual blossoms are termed in this family florets. If we separate one of these florets, we find it is of tubular form, with the mouth cut into five pointed lobes which bend outwards. Cutting away one side of the tube to reveal the internal arrangements, we find five stamens springing from the walls of the tube, the filaments separate, but the anthers are joined together by their sides, so that they form a tube into which the upper end of the style extends. The upper part of the style really consists of two half-round stigmatic arms which are now pressed closely together, their entire surface covered with downy points. At present it serves as a plug to the staminal tube, and as the anthers

Hemp Agrimony.

shed their pollen before the stigmas expand, the tube is filled with pollen. Then the style begins to lengthen, and the unripe stigmas are pushed like a brush up the tube, and the pollen is piled up at the mouth of the flower, and insects crawling over in quest of honey get well dusted with it.

At length the pollen has all been brushed out, and the stigma having passed the summit its two arms separate, so that these catch against insects that have visited an earlier flower and got dusted with its pollen. Some of this is sure to be attached to the stigmas, and the seed-eggs thus cross-fertilised. Fertilisation is effected chiefly by butterflies, which may be found swarming upon it. I have seen plants growing on the face of Cornish cliffs with scores of Painted Ladies, Peacocks, and Red Admirals, the former predominating, crowded upon the flower-heads and drinking the nectar. Bees and flies are also included among its visitors. Self-fertilisation sometimes takes place owing to a few pollen-grains adhering to the hairs of the stigma and getting upon the sensitive surface.

Portion of Daisy-head showing arrangement of florets

Now let us turn to the familiar and ever charming Daisy (*Bellis perennis*), that whitens our lawns and pastures from New Year to Christmas. Surely no European plant beside the grasses is so common. The flower-head—what we mean when we speak of the Daisy—is a more compact, more finished-looking composite than Hemp Agrimony. More than two hundred tubular, bright yellow florets are closely packed together on a conical receptacle, and all these contain both stamens and pistils as in Hemp Agrimony, but the anther cells have got little tails, and the stigma arms are short and thick, covered with fleshy points. The Daisy shows a great advance in the co-operative idea, in the distinction made between the outer row of florets and all the others. This outer row, consisting of about fifty florets, following the example of Guelder Rose, has been adapted for advertising purposes: the corolla has been greatly lengthened, and the yellow turned to white, then it has been split down on one side and the tube flattened out to form a strap-shaped "petal" to the composite flower, but its real character is revealed by the teeth left at the end of the strap to indicate the petals of which the tubular corolla was originally made up. It is really remarkable how, in innumerable instances where the character of a part has been completely changed, Nature appears to have left indications such as these teeth to guide us to a true understanding of the changed organ, yet until recently we have persisted in regarding such marks as meaningless ornament.

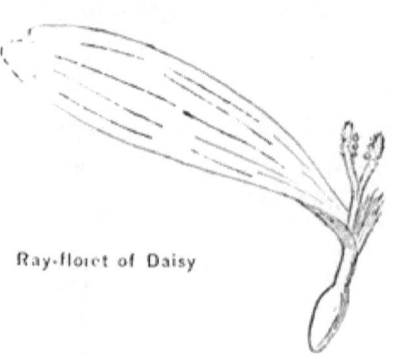
Ray-floret of Daisy

In some species there are three of these teeth (in the Daisy two only, in the Dandelion the full five) at the tip of the ray, with a couple of little petals or long teeth where the ray joins the corolla-tube; in garden Chrysanthemums all stages of transition from disk-florets to ray-florets may be seen. In order to effect this great increase of size in these "ray-florets," as we must term them to distinguish them from the yellow "disk-florets," something has had to go.

It is a common rule in plant-life that display is effected as the result of economies in other directions: every one of these ray-florets has given up the production of pollen, and its stamens have disappeared. It would perhaps be more correct to say that the original head of flowers having taken to the method of developing its anthers before its stigmas, found that the pollen produced by the outer series of florets was wasted, there being no stigmas on that head ripe to receive it; *then*, the material saved was expended in the production of the flag-like rays. They still produce ovaries and pistils, but the material of which stamens and pollen would have been manufactured has gone into the conspicuous rays. There is a distinct difference in the stigmas of the disk and ray-florets; in the former these are short oval lobes, in the latter they are more slender and narrower. The florets are not all of the same age, and so we find that the outer ones discharge their pollen first, and the innermost last. It takes several days to accomplish the expansion and fertilisation of the whole 250 florets, and consequently the Daisy remains fresh for a comparatively long period. Whilst it is there-

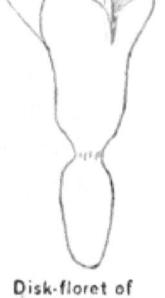

Disk-floret of Daisy

fore certain that a large number of florets will be fertilised by pollen brought from another Daisy plant, quite as many are bound to be affected by the pollen of the more central disk-florets of the same head. By this arrangement, which is pretty general throughout this great family, two advantages are secured: the continual crossing maintains the vigour of the race, and the fertilisation of the older by the younger florets of the same head insures the production of abundance of seeds.

In spite of the earlier conclusions of Darwin, Lubbock, and others, that self-fertilisation is an evil, the enormous number of species of Composites, the abundance of individuals and their world-wide range, convince me that it is the happy mean between continual self-fertilisation and exclusive cross-fertilisation that "pays" best.

The flowers of the Flea-banes, though yellow, are of similar structure to those of the Daisy, but the anther-cells end each in a little tail. The Common Flea-bane (*Pulicaria dysenterica*), figured in our plate, is a familiar example by moist roadsides and ditch-banks.

A botanical description of the flowers of the Chamomiles (*Anthemis*) would agree almost exactly with that of the Daisy; but the disk, which is only a low cone in the latter, is considerably elongated in the former. The plants, moreover, present a striking contrast, such as would never permit confusion, were the flowers even more alike. The Chamomile (*A. nobilis*) forms a branching stemmed, leafy tuft, covered with finely divided, downy leaves, which exhale a sweet and pleasant aromatic fragrance. The smell of the plant is quite the opposite of its

Flea-bane.

taste, for Chamomile has long been popular in medicine on account of its bitter tonic properties, which have also made it valuable as a febrifuge. You cannot walk in the neighbourhood of Chamomile without detecting its presence by the sweet aroma. Yet in the same genus is placed the Stinking Mayweed (*A. cotula*), which, though not differing widely in appearance, is as forbidding as the other is attractive. Its leaves lack the downiness and the fragrance of the Chamomile, and instead are covered with numerous glands from which exudes a nauseous acrid fluid, which not only offends the nostrils, but burns and blisters the hands of those whose business it is to weed the cultivated fields where it grows. The flowers of this Stinking Mayweed differ from those of Chamomile and Daisy in the fact that its ray-florets are usually barren — containing neither stamens nor pistils.

Somewhat similar to the foregoing are the species of *Matricaria*, to one species of which (*M. chamomilla*) the misleading name of Wild Chamomile has been given, because it reproduces in a weaker fashion the pleasing fragrance of *the* Chamomile. They may be at once distinguished from the species of *Anthemis* by the obvious fact that the flower-heads instead of being borne singly at the end of a long stalk are combined into a corymb. This kind of flower grouping is produced by the frequent branching of the flowering stem at its upper part, but the short branches which are also the foot-stalks of the flower-heads are of different lengths, so that the flower-heads are all brought to the same plane. It is the practical result of the umbel attained by a different method.

The other native species is called the Scentless Mayweed (*M. inodora*), because, apart from its clustered flower-heads, in foliage and habit it resembles *A. cotula*, but gives forth neither bad nor good odours. It has been noted that bees appear to dislike the tonic odour of *M. chamomilla*, for very few of them visit it; the work of fertilisation being performed by flies which have distinctly different tastes.

The Chrysanthemum group gives us three native species, one of them suspected of being originally an escape from cultivation that has got itself thoroughly naturalised in places. The two *bonâ-fide* natives are the Corn-marigold (*Chrysanthemum segetum*), with entirely yellow flower-heads, and Ox-eye Daisy (*C. leucanthemum*), with large white rays. The dubiously-indigenous species is the Feverfew (*C. parthenium*). All these differ from *Anthemis* and *Matricaria* in the character of the leaves, which are less deeply and intricately divided, the lobes being broad. Each arm of the style ends in a distinct brush, and before these separate the brushes are side by side, serving to effectually clear the corolla-tube of all pollen. There are some curious variations of this pollen-brush in allied species: thus the Ox-eye Daisy has such a brush in the ray-florets, but as these produce no pollen the brush is not so highly developed. This fact lends force to the supposition that in these Composites the ray-florets did originally possess stamens, but with their suppression the brush-like character has become modified. The reverse of this modification is seen in the garden Marigold (*Calendula officinalis*), in which the disk-florets possess no stigmas, the seeds being produced by the ray-florets alone,

But the pollen of the ray-florets is produced in the manner already described for the Daisy, so the style terminating in a brush is still retained for the sole purpose of pushing the pollen out where it will be accessible to insect-visitors; in this species the style does not separate into two arms.

Mr. F. E. Hulme, in his *Familiar Wild Flowers*, tells us that the name Feverfew "points to the old belief in its efficacy; the significance being that fever patients need be but few in number were the virtues of this plant sufficiently appreciated and utilised." Surely this is a little misleading! The old Latin name of the plant was *Febrifuga*, or the plant that puts fever and ague to flight, and it is more reasonable to suppose that the English folk-name is a mere oral corruption of that word.

The heads of Milfoil, or Yarrow (*Achillea millefolium*), are only about one-fourth of an inch across, but a great number of heads are crowded into each corymb, and the corymbs themselves are bunched together. Owing to this crowding together to produce a more striking effect, the white or pink rays, which are only five or six per head, are roundish or rather broader than long. They produce pistils only, whilst the greyish-yellow disk-florets contain both pistil and stamens. Yarrow is rich in honey, and attracts many insects. The leaves are deeply cut into many pinnate lobes, and these are cut again into five segments, the general effect being delicately feathery. The only other native in the genus is the Sneezewort (*A. ptarmica*), which has slightly larger heads, of which a few only are gathered into a corymb, but this arrangement allows of a larger number of rays, which vary from eight to twelve. The disk-florets are greenish, and the strap-shaped leaves are simply toothed, instead of being deeply divided.

It is worthy of notice that many of the plants we have just been describing have developed strong bitter, tonic, and astringent properties, which cause them to be disliked by browsing animals; even the caterpillars that eat them are few in number. Tansy (*Tanacetum vulgare*) and the Wormwoods (*Artemisia*) share in these bitter properties, and have long been esteemed in medicine for their vermifugal qualities. Tansy is an example of a Composite that has not yet gone in for rays. All the florets are tubular and yellow, but the outer series is so far differentiated that the florets contain pistils only. In the Wormwoods the yellow or reddish heads are very small—about one-sixth of an inch in diameter—but crowded together in panicles or racemes. In spite of this massing, they are very inconspicuous, and do not attract insects: they are fertilised by the wind, and consequently their pollen is smooth—a condition quite exceptional in the pollen of this family.

The closely allied though separated Coltsfoot and Butterbur are interesting as examples of plants that produce flowers long before a leaf appears above ground. Coltsfoot (*Tussilago farfara*) is a well-known weed of stiff soils, in which its rootstock burrows deeply, branching in all directions. About the end of February an unopened drooping flower-head pushes through the earth, supported on a cottony stalk partially covered with oblong scales. The flower-head opens and looks like a large Daisy with

Ragwort.

yellow rays, except that instead of there being but one series of rays as in the Daisy there are several in Coltsfoot. The arms of the style do not separate in this species, but form a rough-headed brush to sweep out the pollen. The seed is crowned with a number of long slender white hairs (*pappus*), which act as a parachute to float it away from the parent plant. Although so beautifully soft to the touch and in appearance, if the individual pappus-hair is examined with a lens it will be found to be rough, with many teeth along its edges. The cobwebby leaves, almost a foot across, make their appearance after the flowers have passed. Large as these are, they are but small by comparison with those of the Butterbur, which are often a yard across. This plant, which affects wetter situations than Coltsfoot, instead of sending up its flower-heads singly, has them all attached to a stout leafy stem. These differ in different plants, one producing flower-heads that are mainly made up of stamen-bearing florets, another consisting of heads that are composed almost entirely of pistillate florets. The staminate heads, however, bear a few pistils, and the pistillate a few stamens. The flowers in this species are flesh-coloured or pale-purple.

The Ragworts (*Senecio*) comprise nine native species, of which several are quite rare. Two of the commonest will serve our purpose just now. The Common Ragwort, or St. James'-wort (*Senecio jacobæa*), grows abundantly in pastures and by country roads, its corymbs of showy golden heads making it very conspicuous. Its leaves are beautiful in form—cut pinnately into lobes, and these more or less cut up or toothed. The arrangement of the florets is the same as in the Daisy: the yellow rays are pistillate, and the disk-florets contain both pistil and stamens. The Water Ragwort (*S. aquaticus*) is very similar, but larger and of laxer growth.

In striking contrast to these is the Common Groundsel (*S. vulgaris*), an annual weed beloved of cage-birds, but detested by the gardener. This plant flowers all the year, like Chickweed. Originally a native of the colder parts of Europe and North Africa, it has gone with colonising Europeans to all the cooler places of the earth. From sixty to eighty florets are packed together in each drooping head, but all are tubular, and consequently the cylindrical heads are not very conspicuous. Occasionally they may be visited by insects, but appear to be principally self-fertilised. In this respect also Groundsel agrees with Chickweed, though it is probably not an instance of degeneration. The leaves are of a simpler character than those of the Ragwort. The fruits are provided with an abundant white pappus, which has obtained the name Senecio for the genus—from *senex*, an old man, the pappus (a word, by the way, which has the same meaning as *senex*) being supposed to represent the white poll of the patriarch. When the seed is ripe, the pappus-hairs separate as widely as possible, forming a fluffy sphere, which may be wafted great distances by the slightest breeze.

An allied species, the Mountain Groundsel (*S. sylvaticus*), shows some advance upon the last-mentioned, from which it has perhaps arisen, for its leaves though similar are more deeply cut, its heads are more numerous, held more horizontally, and often

Carline Thistle.

provided with rays, which, however, are rolled back, and therefore not so very conspicuous. The plant is more or less covered with a glandular down which gives off an offensive odour, and no doubt serves to protect the plant from being browsed down.

A more local plant, the Stinking Groundsel (*S. viscosus*), has these points more accentuated. The heads are larger, with curled rays, and borne erect, and the fœtid odour stronger than in the Mountain Groundsel.

With the bold and striking Burdock (*Arctium lappa*) we seem to get a first suggestion of the Thistle type of flower. The large, thick, heart-shaped leaves, with their woolly under-sides, are suggestive of the Butterbur; but the purple florets, which are all tubular, are enclosed by a many-scaled globose involucre, which gives the resemblance to a Thistle-head. This involucre is an interesting feature of the plant, for the numerous scaly bracts of which it is composed end, each one, in a hard slender point, whose tip is turned down to convert it into a fine hook. The seed-head, therefore, is like a hedgehog whose every spine has been barbed, so that any mammal or bird brushing against the plant will certainly carry off one or more of these prickly burs sticking tightly to fur or feather, and only to be got rid of after much pushing through hedges and thickets, each effort probably resulting in the ejection of a few of the contained seeds.

In the Carline-thistle (*Carlina vulgaris*) we have an advance in thistliness—the leaves being furnished with spines, and some of the involucral bracts being spiny and rigid, to dishearten browsing beasts. The inner bracts are straw-coloured, and of the quality to which the popular term "everlasting" is applied, but the outer ones are leafy with spiny teeth, and between these two kinds is a third set coloured purple. The florets are also purple, and these are all tubular; but the bracts are the most interesting feature of the flower-head, for these have hygroscopic properties, and during dry weather they expand widely. When the air is damp they assume an erect attitude, protecting the florets or the ripening fruits as the case may be. The heads are often gathered for decorative purposes, as they are very persistent in a dry state, and act as weather indicators by the conduct of the bracts. It may not be uninteresting to add that the names of the plant are said to commemorate the fact that Charlemagne and his army were saved from the plague by using this plant as a medicine, after an angel had pointed it out to him! Its healing qualities are said to be found in an acrid resin produced by it. Others contend that the name Carline is suggested by the withered appearance of the flower. See plate 20.

The Knapweeds and Blue-bottle (*Centaurea*) come nearer to the Thistles. There is the same globose involucre, with a large number of closely overlapping scales, in some species with broad marginal appendages, in others spiny or toothed. Here again all the florets are tubular, but the outer ones are greatly enlarged, and the five lobes of the corolla instead of being mere teeth, as in the Daisy group, are here drawn out to a great length, which gives the head a light appearance, when the bulk of the involucre would otherwise make it look heavy. Among the sturdy weeds that take possession of the

Hard-heads.

corners of pastures, and sometimes the very centre thereof, we are almost sure to find the Black Knapweed (*Centaurea nigra*), with its bristly, lance-shaped leaves and purple flowers. The name *Black* Knapweed has reference to the almost black (dark-brown) hue of the comb-like margin to the bracts. Knapweed should be Knop- or Knob- weed, in allusion to the form of the involucre. In this species the outer series of florets is not always larger than the inner ones; but frequently these florets considerably exceed the stature of the others, and on examination they will be found to possess neither stamens nor pistil. In this genus the pollen-brush is not situated at the tips of the pistil arms, as in most of the Composites, but consists of a ring of hairs just below where these branch off. The anthers form a tube as usual, but the folded arms of the pistil are in the centre of that tube, and the pollen is shed in the space between the pistil and the anthers. If no insect-visitor arrives, the arms separate, and some of the pollen-grains fall upon the stigmatic surfaces, and self-fertilisation is effected. But these flowers are much visited by insects for the sake of their honey, and the probability that they will arrive before the stigmas are mature is very great. Müller has observed no less than forty-eight different species of insects on the flowers of the Black Knapweed. The stamens are irritable, and if an insect only touches the tip of one of the anthers, contraction of the whole of the stamens takes place, which has the effect of so exposing the pollen that it is gradually removed by the under-sides of these visitors, who cross-fertilise older blossoms with it.

Hardheads (*C. scabiosa*) is a larger, handsomer

plant, the leaves so deeply lobed as to be almost pinnate, and the heads sometimes as much as three inches across. These are bright purple in hue, varying to pink and white. The outer florets have long rays, and are barren as in Knapweed. All the florets are longer than in that species, consequently the honey cannot be reached by all insects, so it is not surprising that Müller's observations on this plant gave him the names of twenty-one insects as compared with the forty-eight that patronised *C. nigra*.

The favourite Cornflower, or Blue-bottle (*C. cyanus*), has only a few spreading lobes to its slender leaves, and the heads, which are smaller, have a flatter appearance: the bracts have white and brown teeth. The disk-florets are purplish-blue, but the few larger ray-florets are bright blue. Here again the stamens are very contractile. The rare Star Thistle (*C. calcitrapa*), a biennial of dry wastes in the southern half of England, has small rose-purple heads, less than half an inch across, but very singular in appearance on account of the bracts ending in long yellow spines, which give a (conventional) starlike aspect to it. This appearance is more marked in the larger introduced species, *C. solstitialis*.

We make acquaintance with the true Thistles through the medium of the finest native species—the Musk-thistle (*Carduus nutans*), a bold handsome plant, growing erectly for four or five feet, with spiny leaves, and solitary half-round crimson heads, two inches across, which give out a musky odour. The outer bracts of the involucre end in long spines, and before the head is open these spines are connected by little cottony wisps that give the appearance of a spider having utilised them as convenient pegs from which to hang his nets. They are purely vegetable, however, but probably they are as effective as genuine spiders' webs in keeping off or detaining any honey-stealer that is seeking to anticipate the opening for business purposes. All the flowers are tubular, but the upper portion of each is expanded and ends in five long slender lobes. The style arms scarcely separate; they are downy, and beneath them is a ring of hairs which serves for sweeping the pollen out of the tube. The pappus-hairs, or thistle-down, are in this group fine but rough; when their work is completed, they drop off the fruit. It used to be written in books on birds that this thistledown plays an important part in the interior furnishing of the goldfinch's nest. The goldfinch is well known to have a great fondness for Thistle-seeds, and perhaps it would have been excusable on the part of the general public to assume that her nest was lined with the pappus-hairs; but a naturalist should have reflected that autumn is the time for thistledown and spring for the building of nests. It is the down from Groundsel and the woolly webs from Coltsfoot leaves that the goldfinch uses for this purpose.

The Welted Thistle (*C. crispus*), which has a branched stem, produces much smaller heads, of more oval form, and to make up for their defect in size these are produced in bunches — another of the abundant illustrations that insect-fertilised flowers will attain publicity by some means: if the association of a hundred tiny florets into one head will not do it effectually, then the heads must also be laid together.

The Plume Thistles (*Cnicus*) are very similar to the genus *Carduus*, but the pappus-hairs are here formed much like little feathers, and there is a tendency on the part of the heads to consist largely of florets that produce either stamens or pistils only, and to include few perfect florets. The Spear Plume-thistle (*C. lanceolatus*) is a fine example of the genus, growing to a height of four or five feet, with handsome lower leaves, sometimes a foot in length. These larger Thistles, before they begin to develop their stems, form very beautiful leaf-rosettes a foot or so across, of perfect symmetry, the dagger-like spines pointing in every direction.

One species, the Dwarf Plume-thistle (*C. acaulis*), seldom develops its stem at all, but is content to spread its leaf-rosette and to finish this off by producing one central flower-head an inch or two across, of crimson colour, which has a fine but singular appearance in the midst of the dark spiny rosette. I fear the beauty of this plant is quite lost upon the farmer who finds it disfiguring his pastures. A worse pest to the farmer is the Creeping Plume-thistle (*C. arvensis*), which is not content with sending clouds of parachute-borne seeds to spring up all over his and his neighbours' lands, but seeks to take possession of an entire field when once it has got a footing, by sending out underground runners from its rootstock, and though the farmer may prevent seeding by energetically cutting off its heads before flowering, the clump still continues to enlarge. It is a beautiful sight on a sunny morning late in autumn, to see a pasture or piece of waste where this Thistle has taken possession and been allowed to produce its seeds.

Spear Plume-thistle.

The beauty now lies not in the shabby Thistles, but in the flock of goldfinches that feed upon the seeds, and fly up at our approach, flashing the gold and white of their wings in the sunshine. The flower-heads of this and allied species are very popular with insects on account of the liberal provision of nectar in the florets, which is accessible to long- and short- tongued alike. Müller in watching the heads of *C. arvensis* made note of eighty-eight distinct species of insects who thus patronised it.

Several of the Thistles are (or were formerly) used as food, among them the Woolly-headed Thistle (*C. eriophorus*), the young stems and leaves being included in salads or cooked; and the Holy Milk-thistle (*Silybum marianum*), of which the stems were eaten. This fact, that these plants are sufficiently esculent to have attracted man, gives us the clue to the formidable array of spines with which leaves, stems, and flower-heads are alike protected. But for these the browsing animals would improve the Thistles out of existence. As it is, we have introduced one animal that can set even these spines at defiance. The donkey has a mouth and lips sufficiently callous to enjoy such fare.

Respecting the last-mentioned species, the Milk-thistle, we may mention a bit of old-time romance connected with it. The leaves are veined with white, and the old story to account for this singular appearance was to the effect that the Virgin Mary let a drop of her milk fall upon the plant, and as it trickled down the leaf it turned the veins white, as they have remained to this day. And unromantic, matter-of-fact science prevents the legend from dying out by calling the species *marianum*!

There are a few more genera of these Composite plants to which we must refer before leaving the family. There is the Chicory (*Cichorium intybus*), which grows by the wayside nearly all over England, and whose bright blue flowers are so striking in character. Early in the year it may be overlooked as a Dandelion or Hawk-bit, which its leaves somewhat resemble, but later it sends up a grooved and angled stem two or three feet high, and from the axils of the stem-leaves produces its flower-heads with scarcely any foot-stalks. The involucral bracts consist of two series, an inner whorl of long ones, and an outer whorl of short, turned-down ones. All the florets have strap-shaped rays, and these exhibit their origin clearly in the five teeth of the broad tip. The Hawkweeds (*Hieracium*) have flower-heads of very similar construction, the yellow florets being all rayed; but I think it would be more advantageous to consider this type of structure in connection with the larger and ever-abundant Dandelion.

Our forefathers set a higher value upon the Dandelion (*Taraxacum officinale*) than is customary to-day, though they were probably not very sensitive to its æsthetic claims for attention; but to them it was an admirable salad and a valued medicine. It still retains these virtues, but as we now depend upon the greengrocer for our salads and the "patent" quack for our medicines, we know nothing of these things. Still, if we have no eye for the useful, we are not insensible to beauty, and can see it even in common things when it is pointed out to us. The

Chicory

thin-textured leaves are all produced directly from the rootstock. They are boldly cut into angular lobes, with the points directed towards the root, and there is a secondary jagging of their margins into smaller teeth. The largest of these lobes have been fancied to resemble the canine teeth of the lion, whence the common name — derived from French *dent-de-lion*. The flower-heads are borne singly on a pipe-like leafless stalk springing from the centre of the leaf-rosette, and there are nearly a couple of hundred florets in a well-developed head. They are all strap-shaped and yellow; the blunt tips notched into five teeth, hinting that the strap was once composed of five petals.

In the accompanying figure of a single floret picked out from the crowd, all the parts are exhibited. At its base is the ovary containing a single seed-egg, and above this the calyx split up into a large number of hairs which surround the corolla, tubular below and flattened out above. Within the corolla are the five anthers or pollen-bags united by their edges into a tube through which passes the style, branching above into two curled arms — the stigmas. When the flower-head first opened, these arms were not to be seen, for they were quite straight and pressed face to face, low down in the anther-tube. The anthers shed their pollen in this tube above the style, and when the insects began to come after the abundant nectar which lay in the lower part of the corolla the style began to lengthen, so that the pollen

Dandelion floret

was gradually pushed out of the tube against the hairy bodies of the insects, and when these flew off to another head, the pollen went with them, and some of it got swept off by the stigma arms of older florets. That has happened to the one we are examining, for you can see the pollen-grains adhering to it: it lengthened its style until all the pollen was pushed out—a ring of hairs below the arms making a clean sweep of it—then the arms separated and coiled downwards, as shown in the figure, and as they were now sticky, any insect carrying Dandelion-pollen that chanced to brush against them would be sure to leave a few pollen-grains upon it.

As the fertilised seed-egg increases in size the corolla will fall off, the base of the calyx will grow long and slender, the calyx-hairs (*pappus*) will get larger and white, and spread out like a parachute. Then any little breath of air will pull it away from the head, and it will sail away with the fruit containing the ripe seed, and will drop to the ground when the air becomes still or the parachute gets weighted with moisture. The fruit will drop point downwards and enter the soil, and as every movement of the air will move the parachute from side to side, the little barbs at the top of the fruit will work it farther into the soil, until the moisture causes the contained seed to germinate into a new Dandelion plant.

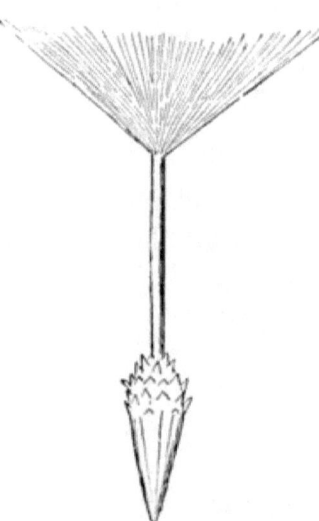

Dandelion seed and pappus

Before leaving the Dandelion, I would add that it is one of the many plants that appear to have a definite period of the day during which the flower-head remains open. This is said to be from 7 a.m. to 5 p.m., and this no doubt has intimate relation to the habits of the insects that visit it. Müller has recorded no less than ninety-three species that come to partake of the abundance of good cheer offered: the nectar rises so high in the corolla-tube that it is generally accessible. The Mouse-ear Hawkweed (*Hieracium pilosella*) is open only from 8 a.m. to 3 p.m., and the Goat's-beard (*Tragopogon pratensis*) opens as early as 4 a.m., and closes eight hours later. This circumstance has given the last-mentioned the alternative name of "John-go-to-bed-at-noon." Like Dandelion, the Goat's-beard has solitary yellow flower-heads, composed entirely of rayed florets. The very long involucral bracts are eight only, and these are joined together at their base. The pappus-hairs are feathered, and of a stiffer character than in Dandelion; the leaves, too, are altogether different, being very long and slender, like grass-leaves.

Closely allied to the Composite family, but separated by botanists, is a small family known as the Dipsaceæ. So far as our native flora is concerned it contains the Teasel and the Scabious only. The principal distinction between these and *the* Composites lies in the condition of the anthers, which are united into a ring in the latter, whilst in Teasel and Scabious they are all free. Again, these plants have a distinct calyx invested by a series of bracts (*involucel*), instead of having it split up into a series of hairs, as shown in Dandelion and other true Composites.

The large lilac-flowered Field Scabious (*Scabiosa arvensis*), the Small Scabious (*S. columbaria*), and the Devil's-bit (*S. succisa*) are our native species. The first-named may be taken as a sample of the three. The head consists of about fifty flowers with four-lobed corollas, of which one lobe is larger than the others, and increasingly so as we proceed from the centre to the circumference. These corollas are of varying length, with wide funnel-like mouths, so that their honey is accessible to short-tongued and long-tongued insects alike; it is consequently a favourite flower with insects of many kinds—bees, flies, butterflies, moths, and beetles, who effect cross-fertilisation. Honey is poured out by the upper part of the ovary, and is protected by the hairy lining of the corolla-tube. The flowers develop gradually, so that the whole head offers attractions to insects for a considerable time, and they are able to come back to the same head day after day and tap fresh stores of honey. The anthers in each flower mature and shed their pollen one at a time, and after the last anther on the *flower-head* is emptied, the styles lengthen and the stigmas all mature simultaneously, so that a single insect that has got well dusted with pollen on an earlier flower-head could effectually cross-fertilise all the flowers on a later head. Some plants produce only flowers with aborted stamens.

Field Scabious and Fruit

The flowers of the Teasel (*Dipsacus sylvestris*) are individually much like those of Scabious, but the head instead of being nearly flat is almost egg-shaped, and the

Teusel.

bracts are rigid, ending in long bristles which prevent insects touching anthers or stigmas with their bodies. The stamens shed their pollen before the stigmas are mature. Humble-bees are the pollen-carriers. The whole plant is interesting on account of its offensive and defensive qualities. Its ancestors have evidently suffered much from browsing animals and from honey-stealers; but at the present time the plant has succeeded in checkmating both. For the herbivorous mammals it has developed a vast number of sharp spines upon its leaves, its many-ridged stems, and the very long involucral bracts. The creeping honey-thieves it not merely keeps away from the flowers on its four-to-eight-feet stems, but it turns the tables upon them so completely as to drown them and then convert them into food. The leaves are in pairs, united by their bases so that each pair forms a capacious basin, with the stem coming through its centre. These basins fill with rain and dew, and crawling insects such as ants, earwigs, caterpillars, and many beetles, in their efforts to reach the flowers, fall into the water and are drowned. You may often find dozens of such victims in a more or less advanced stage of disintegration; and the sensitive cells of the plant have discovered that the water in the basin has become nitrogenous—a kind of Bovril for plants—and have sent out wisps of protoplasm into it, so that its goodness may be absorbed for the general benefit of the entire organism.

Teasel Flower

HAREBELLS

N heaths and pastures where the turf is kept close-cropped, from one end of our country to the other, we may find all summer through one of the most charming of our wild flowers, and one of which poets and prose-romancers of all ages have made good use. This is well known in the South as Harebell, or Hairbell, and in the North as Bluebell (*Campanula rotundifolia*); or, as a compromise between North and South, it may be called the Round-leaved Bell-flower. This last is a book-name, a mere translation of the scientific names. As a matter of fact, it never has really round leaves, but in young specimens the root-leaves are roundish, mostly heart-shaped, and there is a gradual transition from these to the upper stem-leaves, which are exceedingly narrow—the nearest approach to a line, and hence termed linear leaves. From the arched summit of the stem hangs the flower in an inverted position—the proper position for a bell, which the flower closely resembles in form, except that its mouth is cut into five short lobes which curve outwardly.

Years ago it was taught, and with good show of reason, that this bell was so hung on its stem to enable the pollen to fall from the anthers upon the stigmas, whilst the vaulted roof kept out the rain. But this was one of the inferences drawn from a too hasty consideration, without regarding all the facts of the case, and a precisely similar mistake was made in the case of the Fuchsia. Those who made the statement overlooked the important fact that the sticky, stigmatic surface, which is alone sensitive to the pollen, is underneath, and cannot receive the fertilising powder.

Now, this bell-shaped flower tells a similar tale to that told by the florets of the Daisy and Dandelion, as well as by other flowers we have already considered: in order to adapt them better for the visits of insects the five petals have been soldered together by their edges, but the tips of those petals have been left as projecting lobes to serve as developmental clues. As in those Dandelion florets, we find the calyx-lobes are here above the ovary, which is divided into several cells, each containing a number of seed-eggs. There are five stamens springing from the disk, and the base of each filament is flattened out and very broad, the five of them forming a vaulted chamber over the honey-glands around the pistil. The style ends in a club-shaped head, which divides later on into three or five stigmatic arms. A little below this thickened portion the style is covered with hairs.

When the flower opens, the anthers are all pressed against the hairy part of the style, and shed their pollen against it, where it remains entangled in the hairs. Then the anthers start back from the style and shrivel away, and disclose the style muffled up in its coat of pollen. Certain bees, like *Cilissa hemorrhoidalis*, confine their attention to the Harebell, and use the clapper-like style for alighting and climbing up to the honey-glands. Probably they would as often climb up the walls of the corolla, but this is beset with a number of long hairs which poke the bee in the face and so annoy it. An advantage, too, in using the style is that it leads directly to the honey, and so the bee climbs up and gets its under-side well covered with pollen. Several other bees come before the honey is nearly exhausted, and between them carry off the bulk of the pollen. Then the style lengthens, the clubbed head splits into from three to five arms, which spread widely, so that they occupy a considerable portion of the mouth of the bell, and the side of the arms that is exposed outwardly is found to be stigmatic. Should a bee now come, after getting itself dusted with pollen in a younger flower, it will alight on the stigmas and fertilise them.

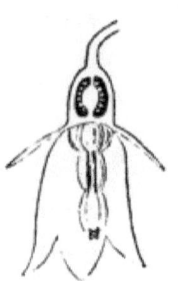
Harebell 1st condition

The Harebell has adopted the hanging attitude for its flowers, doubtless to protect both honey and pollen, but all the native species of *Campanula* are not so careful in this respect. All the others, in fact, bear their flowers erectly, or inclined to the horizontal position, but the plan for fertilisation is the same in all. The commonest of the

Harebell 2nd condition

more erect-flowered species is the Nettle-leaved Bell-flower (*Campanula trachelium*), a plant three or four feet high, that might well be taken for a Nettle before the flowers appear. The leaves are very similar in size and shape to those of the Stinging Nettle, and the bristly stem helps the likeness.

After fertilisation the ovary of the Bell-flowers becomes a seed-capsule more or less shaped like a whip-top, and curiously opening when the seeds are ripe by slits or little doors near the stalk or just under the calyx-lobes. Those of the Harebell and the Nettle-leaved Bell-flower are examples of the first kind, and the Spreading Bell-flower (*C. patula*) is an instance of the second kind.

Most of the Bell-flowers and their allies abound in an acrid, milky juice, which serves to protect them from many animals. Man, however, has found that the acridity may be removed by cooking, so that where there is sufficient substance to warrant the trouble he has utilised some of the species for food. The Rampion (*C. rapunculus*) has a fleshy root, so man has taken it under his care, and cultivated it for eating.

The singular little cornfield weed to which the imposing name of Venus' Looking-glass (*Specularia hybrida*) has been given, is very similar to *Campanula* in essentials, but the corolla is flat, and very small, the lobes deeply divided, lilac without and blue within; whilst the angular ovary and calyx-tube is long and slender, the calyx-lobes being longer than those of the corolla. It has probably become more or less self-fertile, and therefore does not need to produce large flowers for the accommodation of bees or other insects.

The very neat and charming little Ivy-leaved Bell-flower (*Wahlenbergia hederacea*), that grows in some of our bogs, agrees with *Campanula*, except that the somewhat round seed-vessel opens by several valves on its top between the calyx-lobes.

In other genera of this Bell-flower family we get instructive hints of the probable manner in which the five separate petals have become united—in fact, the Harebell itself occasionally explains this to us by producing flowers with the corolla slit into five slender segments, a condition that we find perfectly normal in the flowers of the little Sheep-bit (*Jasione montana*), whose bright-blue heads are usually passed by as those of a small Scabious. A number of florets are included in an involucre, and when separated are found to consist of a top-shaped calyx ending in five long narrow lobes; a corolla split to the base into five slender lobes, which at first are joined by their edges. There are five stamens, the anthers connected by their lower portions and the tips free. The style has a club-shaped extremity marked with the hairy ridges, and developing two short stigmas.

The Round-headed Rampion (*Phyteuma orbiculare*) bears similar heads of deep-blue flowers; but though the corolla is slit into five slender lobes, these are joined together by their edges, though eventually they become free. Its congener, the Spiked Rampion (*P. spicatum*), has the petals permanently attached by their upper portions, but the anthers are free.

The remaining genus, *Lobelia*, of which we have two native species, gives us a remarkable adaptation

Sheep's-bit.

of the same type of structure, with a loss of symmetry. The probability is that the Lobelia flower is derived from a Campanulate form, but the two upper corolla-lobes have become reduced in size and more or less curved back, whilst the lower three form a broad alighting platform for insects. The corolla-tube is split along the upper side to enable insects to get at the honey-glands, the anthers being united into a tube, though their filaments remain separate. The pollen is discharged into this tube much as in the Dandelion, and swept out by a ring of hairs near the end of the pistil, which afterwards emerges from the anther-tube and divides into two short stigma-lobes. All the members of the Bellwort family on being snapped across exhibit a milky juice of an acrid, and in some species poisonous, nature. This fact accounts for the other: that be the turf of the heath cropped ever so closely by sheep or rabbits, the Harebell will be left untouched.

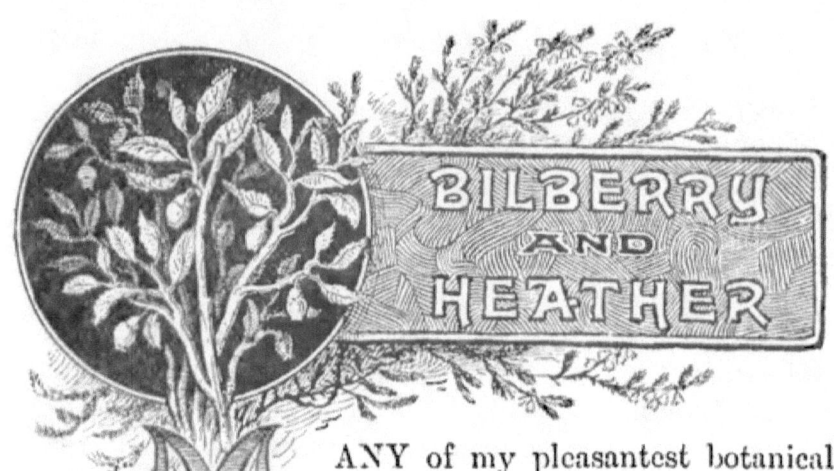

BILBERRY AND HEATHER

MANY of my pleasantest botanical recollections relate to Leith Hill in Surrey, and the delightful valleys of its northern slopes. To lie on a couch of growing heather, doing nothing more than to inhale the balsamic odours from the pine woods all around, and let the eye wander over the glorious expanse of Wealden scenery far below — this, after London, would have been a keen delight; but at a time when the plant-world was opening out its wonders to me, to find myself surrounded here by so many good things was an added joy which the mere sightseer could not share. Without stirring my own length, my eyes, my lens, and my pencil could be occupied for hours by the things within reach of my hands. I could see the singular green and pink caterpillar of the Emperor-moth feeding on the Heath plants, watch the gliding leap of the lizard over the Heather stems, the visits of endless bees to the inexhaustible supplies of nectar in the Heath bottles, and thus be led to examine and sketch the internal arrangements of those bottles and compare them with corresponding parts in the few belated flowers of Bilberry that might still be found, though the "wires" were laden with the bloom-covered berries that served me for refreshment.

Do you know the Bilberry shrub (*Vaccinium myrtillus*), with its angled stems and thick oval leaves, its globular, rosy-green corolla with a small mouth, looking not unlike the nest of a bee (*Eumenes*) I used to find in the Heather close by? Here it shares with Heather the whole of the hill-top not occupied by the pine-trees, and a good deal of the ground under these trees is also covered by it. It is helped in this work of covering extensive areas by the fact that it has a creeping rootstock which runs beneath the ground and sends up innumerable stems. The flower is beautifully shaped, and has a very interesting arrangement of its essential organs. The calyx-tube is top-shaped, and has five short lobes, indicating the five sepals that are amalgamated in it. The corolla, too, has five turned-out lobes round its little mouth, to mark a similar fact in its construction. There is a central thread-like style with a blunt stigmatic tip which stands in the mouth of the corolla for any bee to knock its head against, and round its base are the honey-glands. The stamens are very singular in shape (see figure on next page), and have tubular tips opening at their ex-

Bilberry flower

tremities. From the back of each anther-cell stands out a horn.

When the flower opens, the stamens are all discovered standing with the openings of their anther-tubes pressed against the style, so that the pollen cannot fall out. But should a bee scent the abundant honey and push his head in at the mouth of the flower, and his long tongue go exploring inside, something will happen. First, its face will touch the stigma, and should it already have visited a Bilberry flower some of the pollen it has brought away will be detached. In any case, the poking of the bee's tongue among the stamens is bound to shift several from their position as buttresses to the style, and the moment they are pushed back from their original attitude, the pollen comes out of the anther-tubes and falls upon the bee's face, to fit it for a visit to another flower. Then the ovary develops into a black juicy berry as big as a black currant, covered with a waxy "bloom" which gives a glaucous tinge to it.

Bilberry stamen

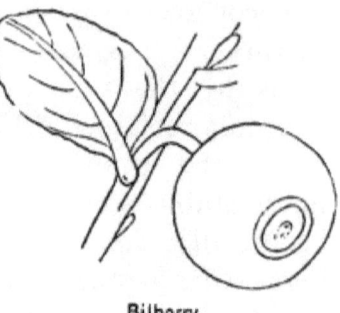

Bilberry

In July and August, as soon as the children are free from school, they troop from the hamlets around, and work among the wires, filling their cans with the luscious fruit ("Hurts"), which makes splendid pies and preserves.

There is some little interest in the names of this plant. Vaccinium is the name by which classical writers refer to some plant the identity of which has been completely lost; but it has been bestowed in more modern times upon this genus in the belief that it was originally spelled Baccinium, signifying a plant that produced berries. Myrtillus has reference to the superficial likeness between this plant and Myrtle, and I think there can be little doubt that Whortle-berry was once Myrtle-berry; yet when one hears these Surrey villagers speaking of the berries as "Hurts," the connection of that word with Myrtle is not immediately obvious. Bilberry was probably once either Bell-berry or Hill-berry; and Blaeberry is, of course, the Northern pronunciation of Blueberry.

The Bog Whortle-berry (*V. uliginosum*) is a similar plant, but of less erect growth, and with smaller berries; found only in the bogs and copses of the mountain districts. The Cowberry (*V. vitis-idæa*) is also a mountain species, but is found more on elevated heathy lands rather than in bogs. Its leaves are evergreen, and its flowers and fruit are clustered instead of being solitary; the anthers, too, are without the horns on their backs. The berries are red, and of a more acid quality than those of Bilberry. One other member of the genus is the well-known Cranberry (*V. oxycoccus*), a creeping plant of sphagnum bogs, with evergreen leaves and a corolla split into four narrow lobes. Its anthers, which have no horns, are yellow, and stand conspicuously in view. The dark-red berries, which are exceedingly acid, are in much request on account of their well-known anti-scorbutic properties.

The beautiful Strawberry-tree (*Arbutus unedo*) is unfortunately restricted as a true native to certain parts of Ireland (its entire distribution being limited to Southern France, Spain, and the Mediterranean region), though it may be seen growing in many woods and parks where it has been planted for ornamental purposes. It is an evergreen, and bears its creamy flowers in drooping clusters. There is a remarkable feature of the stamens to which attention should be called. These are somewhat similar to those of Bilberry, with a couple of horns at the back, but the pores of the anthers are at their base instead of being prolonged into terminal tubes. Between the two anthers there is a projecting tip—a continuation of the filament—and before the flower opens, all the ten stamens are glued to the style by means of these tips. But before the pollen is ripe for dispersal, the attachment of these tips is broken, and the anthers are doubled over on the filament, so that their former base becomes pressed against the style. An insect visiting the flower first touches its head against the stigma, and if it brings any pollen with it from another Arbutus flower, will certainly fertilise this one; then its tongue reaching to the honey-glands at the base of the flower is sure to strike against some of the twenty anther-horns that radiate from the centre, and so shake the anthers that pollen will pour out of the openings upon the insect's face.

It is worthy of notice that Arbutus and all other genera of the Heathwort family have the ovary contained within the corolla instead of below it, as in Vaccinium. That of Arbutus develops into a large succulent fruit, of orange-scarlet colour, and in form much like a round strawberry, and covered with little points. It has not the fine flavour of the strawberry, but when perfectly ripe is a very desirable fruit, especially when it is considered that October and November are the months during which it may be found fit for eating. The Latin name *unedo* signifies that to eat one is enough, the inference being that the inviting appearance is not borne out by the taste, but those who have found it a regular item of dessert in Southern France declare that the name is misapplied.

The Bearberry (*Arctostaphylos uva-ursi*) is very similar to the Arbutus in the form of its flowers, though the cells of the ovary and the ovules differ from it in number, but whilst that is a small tree, this is a depressed, trailing, evergreen shrub, growing on rocky heaths and moors in the North, where its berries are largely patronised by grouse. In other lands they are said to be favourite morsels with bears, and all the names of the plant seek to impress this upon us. The flesh is dry and mealy, and therefore not of a character to commend itself for human consumption.

The true Heaths are contained in the genus Erica, and the most abundant example is the Purple Heath (*Erica cinerea*), which gives its fine colour to moorland and mountain. This has smooth leaves arranged three in a whorl on the stem, and the egg-shaped flowers are also in many whorls one above the other. The Cross-leaved Heath (*E. tetralix*) is as widely distributed, though not occurring in such extensive masses, but its larger flowers make it a better subject for description. In all essential particulars the flowers are like those of the Purple Heath, but the whole plant is downy, and the leaves are in whorls of four arranged crosswise. The delicately rose-tinted blossoms are grouped in a spreading head from the extremity of the stem, and are all drooping. There are four sepals, and four lobes to the mouth of the corolla. Within, the organs are arranged much as in the case of Arbutus, but the stamens are eight in number, the anthers tailed and opening by pores at their sides near the tip. As they stand around the pistil the pores of one stamen are closed by pressure against its neighbours, and the sixteen anther-tails radiate after the manner of the horns in Arbutus, and get in the way of the bee's tongue that is seeking the nectar-glands. Pressure against any one of these tails breaks the ring, and a number of the anthers are able to discharge their pollen. The stigma occupies the mouth of the corolla as in Arbutus, and both anthers and stigma mature simultaneously. In all the Heaths the corolla is persistent—that is, it does not fall off when fertilisation has been effected, nor does it wither up, but it keeps its shape until long after the seeds are ripe, though it loses colour.

The Fringed Heath (*E. ciliaris*) is a species found only in the counties of Cornwall and Dorset. Its leaves and sepals are fringed with glandular hairs, and its crimson flowers are slightly curved and borne in one-sided sprays. The anthers of this and the two following species are without the radiating horns or tails.

The Cornish Heath (*E. vagans*) has leaves like fir-needles, half an inch long, and the flowers are bell-shaped, with an open mouth through which the stamens and pistil are well protruded. The flowers, too, are mounted each on a long foot-stalk, and are

Purple Heath.

massed together in large racemes, comprising from thirty to sixty more or less erect pink flowers. This species is of more restricted range than the Fringed Heath, being confined to the Lizard district in Cornwall, and one or two other localities still farther west.

Our remaining species, the Irish Heath (*E. mediterranea*), is still more local, being in fact regarded as a native of these islands by reason only of its foothold in Mayo and Galway. This, which is a taller-growing species, has a pink corolla of a form half-way between the egg-shape of Cross-leaved Heath and the bell-shape of Cornish Heath.

These Heaths tell an important story, a botanical romance that supports the conclusions of the geologist, and also revives the old legends of Atlantis and Lyonesse, the submerged lands of the Atlantic. The story is too long to be properly told here, but I may briefly say that Plato described Atlantis as an island over 3000 miles long by 250 miles wide, in the western ocean, opposite the Straits of Gibraltar; whilst old chroniclers speak of a land called Lyonesse which connected Land's End with Scilly. Well, geology and botany show that such a land once existed where now are the Azores, connecting not merely Cornwall with the adjacent archipelago, but also with Ireland and the Spanish Peninsula. These Heaths are all found in Spain, and those that are so rare and restricted in range in England and Ireland are very common there, with other species not found here at all; and looking at all the facts of their distribution, which cannot be gone into here, it is concluded that this submerged land was the original home of the

Heaths, from which they made their way to South-west Ireland, South-west England, Western France, Spain, and Portugal; and in the case of the two hardier species, right the way up to Shetland and Northern Europe. The botanical portion of the evidence that this Atlantis existed does not consist solely of the Heaths: there are other plants of similar distribution, including the London Pride and another species of Saxifrage, also the great Irish Spurge, whose strange occurrence in Ireland and the Iberian Peninsula cannot otherwise be explained.

The Heather, or Ling (*Calluna vulgaris*), though popularly associated with the Heaths, differs from them in important points. Its leaves are very small, and overlap each other; and the flowers may easily delude the novice. The sepals are coloured and longer than the bell-shaped corolla, whilst four little bracts under the calyx look like sepals. The eight stamens have horns, and the anthers open by short slits at the side. The flowers hang more or less horizontally, and to allow more room below for bees to get at the honey, the style and anthers take an upward inclination, but the horns secure the shaking down of the pollen upon the insect, and as the style is long and protrudes beyond the sepals it comes first into contact with the visitor's head.

Another of the Heath-like plants whose range is Iberia, Azores, and Ireland, is the St. Dabeoc's Heath (*Dabeocia polifolia*), which is a native with us only by reason of its holding its footing on the boggy heaths of Mayo and Connemara. It has slender oval evergreen leaves, and beautiful rosy, purple, or white flowers, of a pitcher shape.

Similar in the story they have to tell are the Scottish Menziesia (*Phyllodoce cærulea*) and the Trailing Azalea (*Loiseleuria procumbens*), the first found in this country only rarely on the Sow of Atholl in Perthshire; and elsewhere, in the Arctic portions of Europe, the mountains of Western France, the Pyrenees, Northern Asia, and North America. Trailing Azalea is also an alpine plant, whose British range extends only from Ben Lomond to Shetland, at altitudes between 1500 and 3600 feet. Elsewhere this plant occurs only in Arctic and Alpine Europe, Asia, and America; so that these two plants are representatives of quite other climatal conditions than those which induced the Heaths to penetrate northwards — they are remnants of a vegetable invasion from Northern Scandinavia during the Glacial Period, when the cold was sufficiently intense to allow such plants to extend over the old land surfaces as far south as the Pyrenees.

PRIMROSE AND PIMPERNEL

THERE are few flowers whose coming is so eagerly looked for in the spring as those of the Primrose (*Primula vulgaris*), though to speak of it as a spring flower somewhat belies both its popular and scientific names. Primrose and Primula both come from the Latin *primus*, first, in allusion to its early appearance; yet, if we look up the "Floras," we shall find their authors agreed in setting down the flowering-time as April and May. In the South it appears with the Snowdrop and the Daisy, and therefore may be reckoned as one of the first flowers of the year, but until the warm showers of spring come the flowers have not their full beauty or size, and lack the setting of the tender crinkled leaves. The Primrose is one of the plants that arrange for their spring display in the previous summer by laying up great store of material in the thick, fleshy rootstock; and when it puts forth its first flowers the leaves are not one-fourth of their proper size. With the alternating smiles and tears of April the leaves enlarge, the flowers increase in size and number, and their stalks lengthen so that they arch over and look as though they had been lavishly strewn along the copse.

A slight consideration of the flower would convince any of my readers who have followed me through the foregoing chapters that it has been adapted

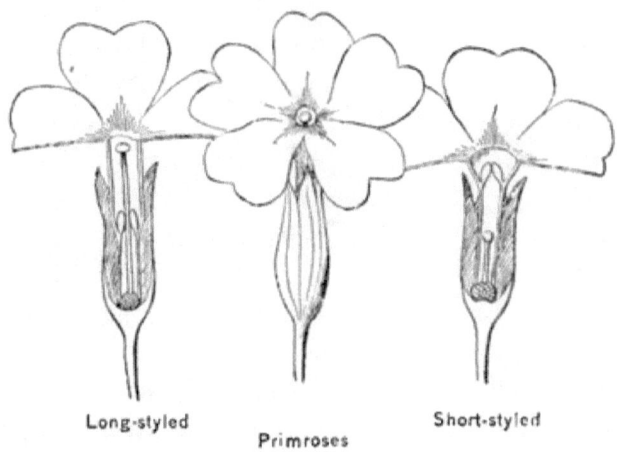

Long-styled　　　　Short-styled
Primroses

specially for the visits of insects with long tongues. The sepals have had their edges joined so that they form a green tube; but to show that this was originally composed of five pieces, the mouth of it ends in five long teeth. The five pale-yellow petals have been similarly united into a slender funnel-shaped tube ending in five broad lobes, each with a central marginal notch. If you will look over the bunch of flowers you have just gathered, you will notice that the mouth of the corolla-tube presents two different aspects in different specimens: one being almost blocked by the thickened head (stigma) of the long style, the other narrowed by the five anthers round the margin. These differences have long been known, but no one guessed their meaning, or thought they had one, until Charles Darwin pointed it out, and now it forms one of the most easily verified examples of the numerous contrivances existing among our wild flowers for effecting cross-fertilisation.

  Wet the small blade of a sharp penknife and with it cut through a Primrose flower, from the stalk upwards; then you will see at the bottom of the corolla-tube an egg-shaped body (ovary) ending in the slender style with a thickened head (stigma). Around the ovary a small quantity of nectar is produced, and only insects with long tongues can reach down to it. If the stigma of the flower you have cut comes to the top of the tube, look out for another in which the anthers are just inside the mouth, and make a section of that in the same manner. Then take a bristle, or something equally slender, and, pretending it is a bee's tongue, pass it down the tube from above. If the flower you first experiment with be a long-styled one, the bristle will be covered with pollen from the anthers which are about half-way down the tube. Then use the same bristle with the short-styled flower, and you will find that the pollen from the first is now detached by the viscid stigma of the second specimen, which is at the same height as are the anthers in the first. At the same time, pollen from No. 2 is deposited on the bristle at a height corresponding with the stigma of No. 1. There is a difference also in the size of the pollen-grains corresponding with the length of the

Primrose.

style they have to penetrate, those of the short-styled form being much larger than those of the other form. By these peculiarities in-breeding among Primroses is effectually prevented, for the plant that produces long-styled flowers does not present us with a single short-styled blossom, nor does the plant with the short-styled produce a single long-styled flower.

There are no less than five native species of Primula: the Primrose, the Oxlip, the Cowslip, the Bird's-eye, and the Scottish Primrose. With the exception of the last-named, all these agree in having the long- and short- styled flowers—or, as modern botanists say, they are *dimorphic*—and it therefore seems probable that the ancestral form from which these existing species were evolved had become dimorphic before these branched off.

These plants all have the flowers grouped in an umbel which is borne on a tall leafless stalk, with the apparent exception of the Common Primrose, which, everybody knows, produces a cluster of flowers, each on a long slender foot-stalk from the heart of the leaf-rosette. But a vertical section through the entire plant would show that even the Primrose flowers are arranged like those of its congeners, only the common stalk is so stunted, or suppressed, that the flowers appear to spring singly from the rootstock direct. Sometimes, however, the Primrose resolves to let us see the true nature of its flower-grouping, and it develops the common stalk, so that it closely resembles the Oxlip (*Primula elatior*).

The Cowslip (*P. veris*) has deeper yellow flowers, of smaller size, and with less inflated calyx; and all three of these have wrinkled leaves. The Bird's-eye

Primrose (*P. farinosa*) has leaves smooth above and covered on the under-side with a meal-like excretion of wax; the flowers are lilac, with a yellow mouth to the tube.

The Scottish Primrose (*P. scotica*) is similar, but smaller, the flowers blue-purple, and the plant occurring only in Orkney, Caithness, and Sutherland; whilst the Bird's-eye's only claim to be regarded as Scottish is its occurrence at Peebles, then from the Border it extends southwards as far as Yorkshire and Lancashire. The small size of the Scottish Primrose must be considered in connection with its loss of the dimorphic condition which marks the other native species: it is not so strongly insistent upon cross-fertilisation, yet those who have studied it tell us that it is chiefly so fertilised.

Before leaving the Primroses, we may note that pollen-collecting bees get no good from their visits to the long-styled flowers. The stigma blocks the entrance, and the anthers are far below; but with the short-styled form the pollen lies most accessibly at the mouth of the corolla-tube, and in getting at it the bees often effect self-fertilisation by shaking some of the pollen down upon the short-styled stigma. Hermann Müller speaking of the behaviour of a pollen-collecting bee (*Andrena*) on the short-styled Oxlip, says: "It holds the anthers in the mouth of the flower in its fore-feet, bites the pollen loose with its mandibles, and sweeps it with the tarsal brushes of the mid-legs into the collecting-hairs of the hind-legs. It visits the long-styled form also, but flies away immediately; not, however, without performing cross-fertilisation in the momentary visit. I have never seen a pollen-collecting humble-bee alight on a long-styled flower; it seems to recognise them at some distance and to avoid them." These humble-bees, in truth, are exceedingly clever, for they often act towards Primroses as they do to Clovers and other plants with narrow nectaries: instead of entering the flower in the legitimate way, they approach it from beneath and bite through the nectary near the honey-glands, and so get their wages without earning them by fertilising the plant.

The Loosestrifes (*Lysimachia*) are a small group of diverse species, two being of erect growth and two prostrate trailers, all with yellow flowers. The Yellow Loosestrife (*L. vulgaris*), so called to distinguish it from the Purple Loosestrife (*Lythrum salicaria*), has two distinct forms of flowers, the short-styled form having two short stamens which fertilise in the absence of insects. The five petals are connected at their base, and dotted with orange within. The five stamens are united below into a tube, and attached to the corolla; apparently there is no production of honey. The seed-capsule is round, and the upper portion splits into five triangular segments to discharge the seeds. Of the trailers the best known is the Creeping Jenny, or Moneywort (*L. nummularia*), which is commonly cultivated in gardens, and is said never to produce seed in this country. This would appear to indicate that it is not a true native but an introduction from the Continent, fertilised there by an insect that does not exist in England. On the other hand, it may be that the plant having taken to rooting from the joints of the stem as it runs along the moist earth, has given up the production of seed here as an unnecessary method of propagation, though it still continues to produce flowers. A very similar species, the Woodland Loosestrife (*L. nemorum*) is often known as the Yellow Pimpernel, from its general habit resembling that of Anagallis.

A little-known, though exceedingly plentiful member of this family is the Sea Milkwort (*Glaux maritima*), a seaside plant that curiously reappears in the inland salt districts of Worcester and Staffordshire. On the seashore it grows just above high-water mark, and forces its fleshy root-fibres between the layers of rock. Its smooth, fleshy leaves are covered with minute pits. There is no corolla, but the bell-shaped calyx is flesh-coloured and minutely dotted with crimson. The five stamens arise from the base of the egg-shaped ovary, and their filaments are coloured deep crimson. The style is simple, with little glutinous points at the tip. At first the tips of the calyx-lobes separate but slightly, leaving a narrow entrance to the flower, and at its mouth stands the slightly curved style, with its stigma mature and ready to receive pollen from an insect that has visited an older flower. At present the stamens are so short that the unopened anthers are not near the mouth of the flower. As the flower opens more fully, the calyx-lobes curve outward, so that the now lengthened filaments elevate the anthers above the mouth of the flower, where they would come in contact with a visitor's head, which would take away some of the liberated pollen. I cannot find any special honey-

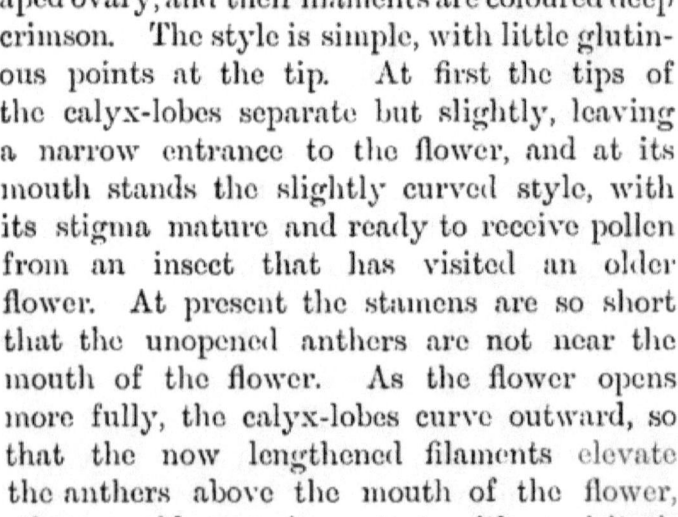

Glaux opening

glands, but the ovary and base of the style secrete drops of liquid, which appear to be accepted as a good substitute. If cross-fertilisation does not take place before the anthers discharge their pollen, there is a possibility of self-fertilisation; but this, I think, is only likely to come about through insect-

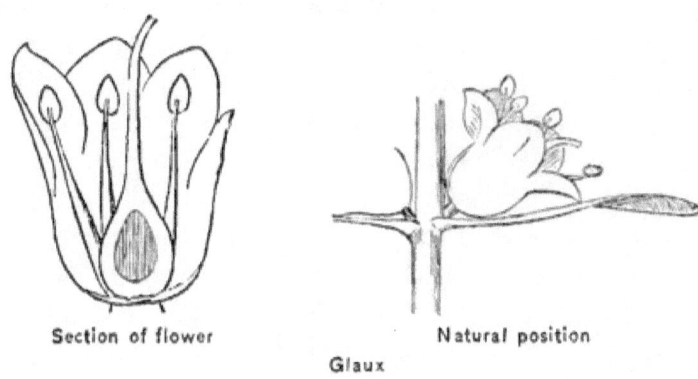

Section of flower   Natural position
Glaux

agency. The Rev. Geo. Henslow has figured the style as curving into a note of interrogation in order to bring the stigma against an open anther, but I believe this was an exceptional example, as in a search extending over four seasons, during which I have examined many hundreds of plants, I have never found a specimen like it.

Somewhat similar are the arrangements of the Pimpernel, or Poor Man's Weather-glass (*Anagallis arvensis*), which has a scarlet corolla. The stamens are attached to the base of the corolla, but the style is only of such length as brings the stigma on a level with the anthers; these in the open flower stand far away from it, so that there shall be no contact, and the style bends between the stamens towards the lower side of the flower, so that it would be the first part touched by an insect-visitor. Yet only an occasional cross can be effected by these tactics, for the flower appears to be seldom visited by insects, offering them no honey in spite of its bright

Pimpernel

hues. At the close of its day it secures the setting of its seeds by the simple expedient of partially closing its petals, which action brings the whole of the anthers into contact with the stigma. It similarly closes up in cloudy weather, and this habit has gotten for it the name of Poor Man's Weather-glass, but unless the poor man is well acquainted with its *penchant* for early closing he may be often misled on this score. It opens at about seven o'clock in the morning, and closes soon after 2 p.m.; so that anyone with faith in its weather-indicating quality, yet ignorant of its business hours, who should refer to it about 3 p. m., when most flowers are open wide, would conclude that the omens were against fine weather. There is a blue variety of the Pimpernel, with smaller flowers, and of more erect habit; but this is rare. Some authors regard it as a distinct species.

The Bog Pimpernel (*A. tenella*) is undoubtedly distinct, for it differs in all its parts. It is a much more delicately built plant, growing on sphagnum moss in boggy places, and producing proportionately large funnel-shaped, rosy-pink flowers with darker veins. It is interesting to observe how in these two species, when fertilisation has taken place, the long slender flower-stalk curls downward so as to hide the swelling seed-vessel under the leaves. When ripe, this fruit is the size and shape of a pea, and instead of opening by five teeth as does the Primrose, it splits in two by a clean horizontal fissure, and the upper portion comes off like a lid.

The last of the Primworts we have space to mention is the beautiful but strangely-named Water Violet (*Hottonia palustris*), which would be more correctly called Water Primrose. The calyx is not a toothed tube as in Primula, but is slit almost to the base into five slender lobes. The lilac corolla is salver-shaped, much like a Primrose, but the lobes longer and narrower.

Water Violet: Long-styled

The stamens and pistils are somewhat similar in arrangement to those of the Primrose, and there are long- and short-styled forms, which are honeyed and borne in several whorls on the tall flower-stem; but, as will be seen in the accompanying figures, the short-styled reaches to the mouth of the tube and the stamens stand around the mouth, whilst in the long-styled form the stigma projects far beyond the mouth. Cross-fertilisation is insured by insects who seek the honey touching the same parts of their bodies alternately against anthers and stigmas; but here, as in Oxlip, mere pollen-seekers (flies in this

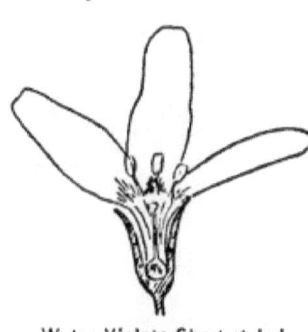

Water Violet: Short-styled

case) bring about self-fertilisation by making several consecutive visits to the long-styled flower, though probably this does not happen to any great extent, for the pollen is very accessible in both forms. Good seed is also produced by flowers that never open, as in the Violet and Wood Sorrel. The fruit is similar to those of Primrose and Pimpernel, but instead of opening as in either of these, it splits from *near* the top to *near* the bottom into five valves, but these valves remain connected at either pole, and only allow the seeds to sift out between. This plant has the alternative name of Featherfoil, which is appropriate enough, for its leaves are cut from each side near the midrib into a great number of slender lobes, which give a very feather-like appearance to the leaves. We have previously seen how the submerging of a leaf causes it to become either entirely slender, or if fairly broad, cut into a number of slender lobes or filaments that will offer little opposition to movements of the water, and at the same time expose a larger surface to catch the small amount of carbonic acid gas that is dissolved in the water.

GENTIAN AND BOGBEAN

EVERY alpine climber is inclined to wax eloquent concerning the beauty of the Gentians "found growing in profusion on little swards environed on all sides by ice-clad rocks and mighty glaciers." Some of those alpine tourists are ignorant of the fact that we have no less than five species of these Gentians growing at home, though they have not the impressive surroundings of the Alps; but to show their partiality for such environment two of these restrict themselves to such alpine situations as our islands afford. The family is noted for its bitter and tonic properties, developed like the acridity of the Buttercups to protect them in large measure, if not entirely, from the attacks of herbivorous animals. In some species this bitterness becomes emetic, and even narcotic.

The Field Gentian (*Gentiana campestris*) has an almost cylindrical pale-lilac corolla with a four-lobed mouth large enough to give entrance to the humble-bees, by which it is largely fertilised; but flies and other uninvited guests are kept out by a fringe of long hairs above the level of the anthers, which

Field Gentian

closes the opening against weak insects. The two-lobed stigma matures slightly in advance of the anthers, and on receiving pollen from another flower the lobes close up, so as to leave the way more open to the anthers, that their pollen may be taken to younger flowers. The similar Felwort (*G. amarella*) has five-lobed pale-purple flowers. They are both adapted for fertilisation by bees and butter-flies, and as they grow among grass—the former in moist pastures, the latter in dry pastures and on downs—the lobes of the flower are spread out to make it more conspicuous from above.

The foregoing are annuals, but the Marsh Gentian (*G. pneumonanthe*) is a perennial, found locally on marshy heaths. Its corolla, though slender, is more bell-shaped than the foregoing, greenish on the outside, but lined with bright blue. There is no forbidding fringe of hairs as in the previous species, simply because such a provision is unnecessary with the other special arrangements of the interior. About half-way down, the corolla becomes abruptly narrow, and to this point a bee can creep. Here it finds the anthers pressed round the style, and gets its under-side powdered with pollen. It is still half an inch away from the honey, and unless its tongue be that length cannot reach it: it is therefore only certain bees, butterflies, and moths that find it worth while to attempt to get it: the stamens mature before the stigma, so that if an older flower is next visited by the bee that has succeeded in getting the honey it will be cross-fertilised, for the bee will have now to crawl over the expanded stigma-lobes upon which some of the pollen from its under-side will be detached.

We have two other native species, both very rare, being confined to the mountainous districts of our land. One of these is the Spring Gentian (*G. verna*), with bright-blue salver-shaped solitary flowers, an inch across, yet the stem of the plant is only an inch or two high. The throat of the corolla is partially closed by five split scales which project from between the lobes. Müller says it is adapted for cross-fertilisation by butterflies and moths; the most important visitor being, apparently, the Humming-bird Hawk-moth (*Macroglossa stellatarum*). The remaining species is the Small Alpine Gentian (*Gentiana nivalis*), with very small funnel-shaped flowers, far less conspicuous than the Spring Gentian, and consequently less visited by insects; but in case these should entirely overlook it or otherwise fail to put in an appearance, it has reserved the right, and power, to fertilise itself.

In our country all the Gentians have the petals joined together by their edges to form a tubular corolla, and are more or less blue; but on the Alps there is a well-known Yellow Gentian (*G. lutea*), which produces its flowers in tiers one above the other, and these flowers consist of five petals free almost to the base. Its honey is freely accessible to short-lipped insects, and they appear to avail themselves fully of the opportunity. Anthers and stigmas mature simultaneously, and are on the same level, so that if no cross-fertilisation takes place, as happens only occasionally, self-fertilisation may be effected by the stamens erecting themselves by the style. Now, it is believed that the ancestral Gentian had open flowers like this and of the same colour, offering honey and pollen to all comers, and getting an occasional cross in return for its hospitality. From this stage evolution proceeded along two lines, in both making the honey less accessible by connecting the petals throughout a gradually increasing part of their length, and so bringing about the prevailing bell-shape, as most befitting the humble-bees who were the principal selective agents. Along one line the honey got to be hidden in deep narrow passages only accessible to those with long tongues, and to secure cross-fertilisation the anthers took up their position round the style, where they must be crawled over by insects that would have honey. And in the variation of the flower-colours those that developed any tinge of blue—even in spots, as some of the Continental forms still possess it—would be selected by preference by bees who have special fondness for that colour.

Our own species, *pneumonanthe*, is an illustration; its blueness is chiefly within. Then some of these got taken in hand by the butterflies, and further modified accordingly, the tube being narrowed and lengthened, and the bilobed stigma developed into a disk partially closing the mouth of the corolla. *G. verna* and *G. nivalis* exemplify this type in our country. The other series developed along the line of excluding unprofitable visitors by hairs on the corolla, but narrowing of the corolla had to proceed at the same time; finally we find, as in our *G. campestris* and *G. amarella*, the throat-hairs and the narrow corolla together exclude all but bees and lepidoptera whose tongues are long enough to reach the honey and at the same time effect cross-fertilisation owing to the position of the essential organs.

There is a pretty plant called Yellow-wort (*Chlora perfoliata*), that grows on downs, banks, and heaths, on a chalky or clay soil, of which perhaps the most striking feature is the perfect amalgamation of a pair of leaves by their lower edges (hence the name *perfoliata*, signifying that the stem passes *through the leaf*). The yellow flowers, however, are interesting, and may help to throw some light upon the original form of the blossoms in the family. It is clearly distinct from our Gentians, yet it is not without suggestions of that Yellow Gentian to which I have referred. Its corolla is tubular at its base only, and above is split up into eight or six spreading lobes; there is an equal number of very slender sepals, and stamens to correspond. The stigma is cloven into two lobes, and to favour cross-fertilisation these mature before the pollen is shed; if, however, insect-visitors fail to arrive in time, self-fertilisation may be effected, but however fertilisation comes about, when completed the style is shed. It may be noted in passing that the object of the perfoliate leaves is to make it difficult for undesirable crawling insects to obtain access to the flowers by climbing the stem.

A more noticeable plant by roadside and in dry pastures is the neat little Centaury (*Erythræa centaurium*), with pink or red flowers, funnel-shaped, with a long cylindrical tube and four or five spreading lobes. There is a good deal of variation in the length of the tube, the proportionate length of the style, and the size of the pollen-grains, much as we found to be the case among the Primroses. The stamens are attached to the corolla as in the Gentians, but these differ in the fact that the anthers in discharging their pollen twist up into a spiral. The flowers do not appear to produce any honey, yet they are visited by butterflies and moths to whose long trunks the form of corolla-tube is adapted; they probably suck juices from the flower-tissue, as is done in Orchids and some other flowers. H. Müller records the Humming-bird Hawk-moth, the Silver-Y moth, and the Yellow-Underwing among its visitors.

More distinctly beautiful is the Bogbean, or Marsh Trefoil (*Menyanthes trifoliata*), which grows on the margins of woodland tarns and in boggy places. This bears stouter funnel-shaped flowers in racemes on a long stalk, the corolla pink without and white within, the inner surface of the five lobes fringed with fleshy filaments that serve to keep out creeping insects. Cross-fertilisation is insured by the flowers being of two forms: long-styled and short-styled.

All the members of the Gentian family abound in a bitter principle, which has no doubt been developed for the protection of the plants against the browsing quadrupeds.

BUGLOSS AND SCORPION GRASS

IN waste places, wherever the soil is fairly light, we may look for a rough leafy plant, two or three feet in height, with curving sprays of red and blue flowers and buds. These sprays are at first short, with the buds closely packed together, but as they open in turn, and become blue, the cyme increases greatly in length and the seed-vessels become distinct one from another. This is the curiously-named Viper's Bugloss (*Echium vulgare*), the second word having reference to the shape and roughness of the leaves, which are supposed to resemble the tongue of the ox—Greek, *bous*, ox, and *glossa*, tongue. The irregularly five-lobed corolla is funnel-shaped, expanding widely at the mouth, and is thus suited for the visits of many insects that differ considerably in size and other respects. The large humble-bees can get their heads well in, and reach the honey with their long tongues, whilst the smaller bees and wasps with short tongues can get well down near the honey, which is secreted by the base of the ovary; and all can do this with such ease and rapidity that a large number of flowers can be treated without loss of time. The plant is therefore a great favourite with many kinds of insects: H. Müller has recorded eighty species from his own observations. The special mechanism adopted to take proper advantage of this popularity consists in hanging the flower horizontally, and bringing the four long stamens all to the lower side of the corolla, so that as they project beyond the mouth they form a convenient platform upon which the insects alight, and the larger ones, such as humble-bees, must get dusted with pollen on their under-sides. The flower has five stamens altogether, but one of these is short, and does not extend outside the mouth of the corolla. This short stamen is a special provision for the smaller bees that fly in without alighting on the external platform. All the anthers as they are about to discharge their pollen turn their faces upwards, to insure contact with the visitors. At this stage the style is short and the stigmas immature, but when the pollen is shed the

Viper's Bugloss

style lengthens until it is longer than the long stamens, the free end divides into two branches, and the tips of these branches are the stigmas. It lies beyond and above the anthers, so that no bee, large or small, can enter the flower without touching it and pollinating the stigmas, if it has previously visited a younger flower. Müller found that two species of bees (*Osmia*) get their own food, and that for their progeny, exclusively from this flower.

Borage (*Borago officinalis*), though not a native, is frequently found in waste places where garden refuse has been thrown at some time or other, and it is worthy of notice because the mechanism for fertilisation is so similar to that of the Violet. The plant is more densely clothed with short hairs than Viper's Bugloss, the object being to repel slugs and other depredators, who would soon destroy a plant with so succulent a stem. The nodding flowers have a brief corolla-tube, from which radiate the five somewhat triangular and intensely bright-blue petals. The stamens are dark purple, with stout filaments, a series of notched scales at their base closing the tube. Each anther has a spur upon its back, and the tips press against each other, thus forming a hollow cone in which is the style. The anthers mature before the stigma, but the pollen is not all ripened at once, so that a succession of bees may be loaded with pollen before the stigma is ready. Owing to the space between the filaments and the corolla-tube being stopped by the scales, the insect must get its tongue between two stamens, which being moved out of their place, let a little shower of pollen fall upon him. When all the pollen has been thus used up, the style lengthens, and the knobbed stigma emerges beyond the tip of the anther-cone, and is then mature. From the hanging position of the flowers, only insects, such as bees, having the power to hang downwards on the petals whilst they thrust their long tongues between the stamens, can obtain the honey.

Prickly Comfrey (*Symphytum officinale*) is another of these bristly herbs whose flowers assume the pendulous position. The corolla is tubular, enlarging towards the mouth, which is closed by spear-shaped, toothed scales, and finished with five turned-out teeth. The five stamens are attached to the corolla-tube, alternating with the spear-shaped scales; the anthers converging to a point, as in Borage, forming a hollow cone into which the pollen is shed before the flower opens. The style ends in a rounded head (stigma) which stands beyond the anthers, and comes to maturity soon after the flower opens. An insect comes into contact with the stigma on first alighting, and if it has already visited a Comfrey flower and got dusted with pollen, cross-fertilisation is first effected, and then by pushing its tongue between the anthers, it shakes down a shower of pollen with which to fertilise another flower. Some of the bees that visit the Comfrey protest against these arrangements by biting lobes in the basal portion of the tube and extracting the honey without touching either stigma or anthers! Such is the stage of intelligence to which these creatures have now arrived.

The particular form of cyme in which the flowers of Comfrey are arranged is known as a scorpioid cyme, because its curve suggests the curl in the tail of the scorpion! A similar form is found in the

Comfrey.

Common Bugloss (*Anchusa arvensis*), another bristly subject growing in fields and wastes where the soil is light; but here the scorpioid character is less evident. The blue tubular corolla is curved, and at the mouth there are five spreading lobes, each furnished with a white hairy scale at its base, and these scales partially close the mouth of the tube. Their white colour also makes them serviceable as honey-guides. Anthers and stigma mature together, but the position of the stigma above the anthers favours cross-fertilisation when insects pay their visits; should these not take place, self-fertilisation may be brought about by the corolla with the attached stamens dropping off, and so dragging the anthers over the stigma.

The Forget-me-nots, or Scorpion-grasses (*Myosotis*), are represented in our flora by half a dozen species, of which *M. palustris* is the true Forget-me-not, and has much larger flowers than the others. In all the species there is a bell-shaped calyx with five teeth, and a salver-shaped corolla to whose tube the anthers are attached much as in the Primrose, whilst the central style also resembles that of the Primrose in having a globular stigma. The mouth of the corolla-tube is thickened, and bordered with yellow, from which five bands of white radiate across the exquisite blue of the corolla-lobes, and serve as honey-guides. Honey is produced at the base of the ovary, and insects have, in order to reach it, to push their tongues between the anthers and the stigma, and so narrow is the space that on one side the tongue touches the anthers and on the other the stigma. In going from plant to plant they are sure to get both sides of the tongue dusted with pollen, and so leave a little on every stigma they touch. This is not a very precise method, because it will produce self-fertilisation if an insect happens to visit one flower twice; but then the Forget-me-nots are not dead-set against self-fertilisation, for there is always a chance of pollen being shaken down from the anthers to the stigma. In *M. versicolor*, if fertilisation has not been effected by other methods, there is still a certainty that it will take place, under the following conditions. When the flower opens, it is yellow, and the style is so long that the stigma stands out beyond the mouth of the corolla, but afterwards the corolla becomes blue, and lengthens until the anthers are brought up to the level of the stigma.

In the Wood Forget-me-not (*M. sylvatica*) the style is much shorter than the corolla-tube, so that pollen may be shaken down from the anthers just inside the mouth of the tube; but it is much visited by flies and bees, and crosses must be frequent. The Field Forget-me-not (*M. arvensis*) is similar, but the flowers are smaller, and the stigma is on a level with the anthers. Four of the native species grow upon solid ground where creeping insects are plentiful, and they are all more or less hairy; the calyx, too, is covered with stiff hooked hairs. The reason for these is obvious when the four little nutlets which constitute the fruits of the Forget-me-nots are ripe. These are still protected by the calyx though the corolla has long departed, and they are usually highly polished, so that they may easily slip through vegetation and into chinks in the soil. But they want to be carried from

Fruiting calyx of Forget-me-not

the parent plant, and if you will walk where the terrestrial species grow you may speedily find out how they are carried: the bottoms of your trousers, or hem of your skirt in the case of a lady, will be found thickly studded with the calyces of these plants. They similarly catch in the coats of mice, shrews, rabbits, weasels, birds, and around the legs of sheep, dogs, and foxes, which shake out the nutlets as they walk or run.

The two species, *M. palustris* and *M. cæspitosa*, which grow half-submerged in streams and ponds, present such a contrast as proves conclusively what is the reason for hairy stems and leaves and hooked calyces. There are a few hairs on the stems, but the leaves are so smooth that they shine, and the calyx is furnished only with straight simple hairs which lie close against it. The reason for the difference is obvious: ants cannot steal the honey or pollen if the plant is surrounded by water, though this may be the thinnest layer possible, and hooks on the calyx are unnecessary where the water is the best carrier of the seeds which are not wanted to be scattered on land.

FOXGLOVE AND TOADFLAX

ONE of the most imposing and stately of our native plants is the Foxglove (*Digitalis purpurea*), especially when this is seen growing in hundreds over some sloping bank or up the face of an escarpment. These are really its favourite haunts, for it is fond of light, well-drained soils; and the flower-lover who comes upon such a Foxglove show will not feel that his country is poor in flowers of striking appearance. Most persons gazing on such a sight would probably be astonished to learn that it is to the humble-bees we are chiefly indebted for the display. It is for their accommodation that these flowers have their peculiar form, the drooping attitude, and the strange position of the stamens and pistil.

We cannot long stand before a clump of Foxgloves on a bright day without hearing a contented buzzing, and seeing a burly black-and-orange-banded humble-bee flying from flower to flower, commencing at the

Foxglove.

lowest on the spike, and in turn visiting all those that
are open. These bees are not mere gad-abouts, intent
solely upon refreshments; they are the supervising
genii who look after the best interests of the Foxglove
what time they are gathering pollen for their family
and a little nectar for themselves. Humble-bees of
several species have had much to do with the moulding of the Foxglove's flower, and all its internal
arrangements have reference to their visits. No
other creature can now perform the offices they serve,
so that the least the flower can do is to keep a little
store of nectar for them in a suitable and not too
accessible place.

Here comes another great fellow, buzzing in a tone
that appears to denote good humour. Watch him.
He commences at the lowest flower in a spike, crawls
right in, and disappears from view. He is out again
in a few seconds, and instead of flying off to another
plant, enters the flower next above it on the same
stem, and so on right up the spike, until he reaches
the unopened flowers. Then he booms off again, and
commences another spike a little farther on, proceeding again from the lowest to the highest.

To know exactly what he is about, we must open
two or three of these flowers. As they hang one
above the other, we note that the side nearest the
stem is longer than the upper side, and its central
lobe affords a convenient alighting-platform for the
bee. It is dotted with conventional "eye-spots," and
furnished with long erect hairs. Having observed
so much, we have no further need for this half of the
flower; we therefore slice it off, and this reveals to us
the inside of the upper half together with all the

organs as shown in the first of these three diagrams of this part. This is of a flower that has only just opened. The shaded portions above represent the green sepals clasping the narrowed portion of the corolla, within which is the pear-shaped ovary terminating in the long style. Between the ovary and the walls of the tube on either side is a pair of stamens, their filaments curiously curved and twisted in order to bring them into one plane. Their plump purple-dotted yellow anthers are placed more or less

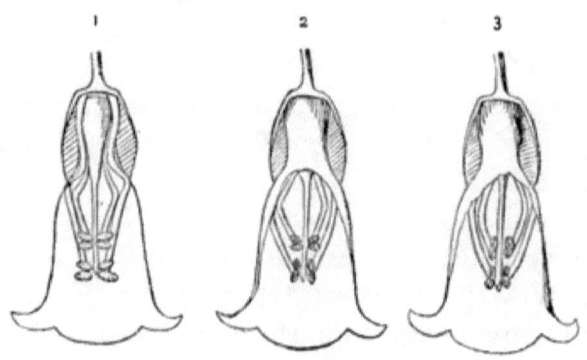

Three conditions of the Foxglove

horizontally, overlapping the style. In this flower all the organs are immature, but if we cut open the flower next below it on the spike, we shall find its condition as shown in the second figure. The anthers nearest the mouth of the flower have assumed a vertical rather than horizontal position, and have burst open, exposing the pollen and dropping some among the hairs of the platform below.

If you will sit down beside the plant, you will be able to catch a glimpse of what happens in the flowers when a visitor comes. There goes a bee. See, he walks over the hairs of the platform, and his hairy under-side picks up much of the pollen, which he will by and by brush off with his feet and add to the big pellets of pollen attached to his hinder legs. That is a bribe to take his attention from the pollen that is being brushed off the anthers by his back as he presses in to suck the honey secreted by the smooth ridge at the base of the ovary. Now look at the flower below, which is consequently older (Fig. 3). All the anthers have assumed the vertical position, have shed their pollen, and are empty. The point of the style appears to have developed a mouth: it has split into two diverging lobes, the stigmas. *Now* the purpose of everything is evident. The anthers have been successively brought into line, their pollen covers the middle of the bee's back, and now the stigmas are mature they occupy practically the same position.

Let us suppose the bee we saw just now in the Foxglove bell has flown to the next plant, well laden with pollen. He commences with the oldest flower, and leaves a little of his pollen upon the ripe stigmas, gets his honey, and goes off to the flower next above it, fertilising that also. At the third flower, perhaps the stigmas are not mature, and in that case he gets more pollen, which can only be used to fertilise the flowers of another plant, because all those above the one he is in will be immature so far as the stigmas are concerned.

You can tell on examining any of these flowers whether the bees have called and done their work; if they have, the anther-cells will be empty, and this is the condition in which we find most flowers that have been open a few hours. But should there be a scarcity of bees, and the pollen be still plentiful in the box-like anthers, self-fertilisation will come about in a singular way. Before the stigmas are mature, you may pull at the Foxglove corolla and have difficulty in removing it from round the ovary; you are more likely to tear it to pieces. But when the stigmas have expanded, the corolla is loose, and after a time slips off; in so doing the anthers are dragged over the stigma-lobes, and any pollen that may have been left is shaken upon them.

When the seed-capsule is fully matured, it splits down each side from the old style to the base, and allows its small light seeds to be shaken out as the long stem oscillates in even slight breezes. Professor Henslow calculates that a single plant of Foxglove produces no less than 1,500,000 of these seeds. That is a pretty liberal provision for the perpetuation of the species.

The Foxglove family is a very extensive one, and even in our own country includes no less than fourteen genera. We cannot deal fully with all these, but will glance briefly at the more striking variations from the Foxglove plan.

The Mulleins (*Verbascum*), before they develop their tall flowering stems, are often mistaken for Foxglove plants, the leaves being of similar dimensions and shape; though in those of the Foxglove the upper surface is rough with wrinkles, those of the Mulleins are in the several species more or less densely woolly. But when both plants are in flower the veriest tyro in matters botanical would never be in danger of confusing them, for the Mullein has yellow flowers of a more open character than those of Foxglove. There

is a tube of some sort to the corolla in all the plants
of this family, but in some genera it is very short.
That of the Mullein is an instance of the latter sort.
At first sight it looks like an open flower with five
distinct petals, but these are united into a brief tube
with five large spreading, nearly equal lobes. It is
very probable that the original founder of the family
had flowers much like Mullein, for its calyx and
corolla both bear plain indications of their fivefold
character—further, they have five stamens, in which
they differ from all the other native genera of the
Foxglove family. The fifth stamen has been got rid
of in most groups, because it came in the way of the
special arrangements made for the insect-helpers.
In one genus, as we shall see, the stamens have been
reduced to two in order the better to utilise the
insects.

The Mullein-flowers hang vertically from the
woolly spike, and whilst the stamens curl in an upward direction the style curves downward or outward,
and is used by some visitors as an alighting-perch.
Anthers and stigmas ripen simultaneously, but owing
to the relative positions of these cross-fertilisation is
favoured, the stigmas being first touched by visitors
—chiefly the smaller bees, with an eye for pollen
rather than honey—therefore the flowers produce a
very fitful and scanty supply of nectar. Should no
visitor do the needful, there is every chance of a little
pollen falling upon the stigmas. The pollen is orange-
red, to enable the bees to find it at once, and the
filaments of some or all of the stamens are covered
with hairs, to enable the bees to take hold more easily
whilst they are gathering the pollen. In some species

these hairs are white, in others purple. The Common Mullein is the *Verbascum thapsus*, the Hæg-taper or Hedge-candle of our Anglo-Saxon ancestors. The old Latin name was *Barbascum*, from the bearded stamens, and this by a process quite common among our forefathers, judging from numerous examples, has been corrupted into *Verbascum*.

The Speedwells (*Veronica*), with their little bright-blue flowers, do not appear at first sight to present any resemblance to the Mulleins, yet apart from size and colour there is not a very wide difference. There is the shallow tube to the corolla in most species, with the stamens attached to its mouth and hanging far out; but, with the reduction in size from the founder of the family the Veronicas have lost one of the lobes of both calyx and corolla; not only so, but they have lost three of the originally five stamens. Supposing that the first Veronica was similar to the common but beautiful Germander Speedwell (*V. chamædrys*), we can understand why only two stamens came ultimately to be developed. The purple style stands a long way out from the flower with a slight upward curve, and the two blue stamens spread out in front of the side lobes. There is a

Germander Speedwell

zone of white round the mouth of the tube—which is blocked to Thrips and other unprofitable creepers by a fringe of fine hairs—and a number of dark-blue lines converge to this from each lobe, indicating the road to the honey, which is secreted at the base of the ovary. Fertilisation is effected chiefly by certain two-winged flies; from the relative positions of anthers and stigma it could not possibly come about without insect aid. The filament of the stamen is relatively stout, but near the base it becomes very thin and flexible; this is intimately connected with the visitors, and the base of the style is thinned to correspond. A fly alights on the lowest lobe of the corolla with the style under it, so that its slight upward curve causes the stigma to be pressed against its lower surface. Then, pressing towards the honey, the visitor by his fore-legs takes hold of the stamens, which bend from the base until the anthers touch his under-side and dust it with the white pollen. The next flower he visits will receive part of that pollen upon its stigma.

It is said by Müller that the anthers and stigma ripen simultaneously; but I have found in half-opened flowers that the stigma is already mature, though the anthers have not yet opened, nor has the honey yet commenced to flow. In this condition the corolla-lobes being rolled one within another, constitute a long tube whose mouth is partially closed by the stigma which will be found, quite mature, in the orifice or extruded to the extent of 1 mm. It is very probable that cross-fertilisation is effected thus early by a pollen-smeared fly trying for entrance or merely walking over the unopened buds before flying off. When the pollen has all been distributed, the corolla is thrown off in a peculiar manner by the calyx, which flattens itself laterally, the pair of lobes on each side becomes erect instead of spreading, and the pressure this double action exerts on the corolla-lobes is such that the base of the corolla-tube is detached and the entire organ is thrown off.

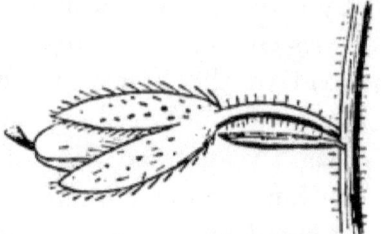

Bursting bud of Speedwell

The object of this is apparently twofold: to prevent waste of time owing to flies operating on flowers already fertilised, and further to protect the ovaries during the development of the seeds. Yet the corolla often flies off before the stigmas have received pollen. Müller mentions the fine hairs that fringe the lower half of the corolla-mouth as serving to protect the honey from rain. Their position and direction do not support this view: such a function would be better served if the hairs projected from the upper lobe of the corolla. But they do effectually block the way to the honey for any weak creeping insects, and it is clear that the plant has to protect itself from such robbers, for stems, leaves, flower-stalks, and calyx are alike covered with hairs to keep them away.

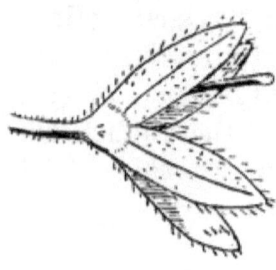

Flattened calyx of Speedwell

The Brookline (*V. beccabunga*), whose stout hollow stems and thick smooth leaves contrast strongly with those of Germander Speedwell, is semi-aquatic, and its freedom from hairs reminds us of the similar differences between the species of Forget-me-nots. Like the Speedwell, it opens with the stigma mature, though the anthers are not. The stamens do not diverge widely, and there is considerable chance of the stigma coming in contact with one or the other, with or without insect aid.

The Spiked Speedwell (*V. spicata*) differs from most other species of the genus in the more tubular character of its corolla, which has a bearded throat and narrow lobes. On some plants the flowers all develop their stigmas before the anthers, in others the style remains dwarfed until the pollen has all been shed; either case favouring cross-fertilisation. This species appears to be chiefly visited by bees; those that we have not mentioned, though visited by insects, are most frequently fertilised by their own pollen.

An advance on the open flowers of Mullein and Speedwell is found in those of the Figwort (*Scrophularia nodosa*), which are somewhat globular, but open in front. The whole plant gives out a fœtid odour, which may often be perceived long before the plant is reached, and which appears to be very attractive to wasps, who visit the livid, red-brown flowers in considerable numbers. The flower-parts are here in the original condition of fives. The calyx is five-parted, and the corolla has five lobes of unequal length; the two longest form an upper lip, and these with a shorter one on either side stand erect, but the shortest of all forms a lower lip which is turned downwards. There are also five stamens, but the fifth is reduced to a mere black scale on the upper wall of the corolla, whilst the four perfect ones lie along the lower wall and extend over the lower lip when the anthers are ripe. The stigmas mature first and project over the lower lip, whilst yet the stamens are short and lie within the corolla.

Wasps on visiting the flower cling to the lower lip and push their heads in to get at the honey secreted by the base of the ovary. If this should happen in an old flower, the under-side of their head and "chest" is covered with pollen from the anthers, and on visiting a younger flower this is transferred to the stigma, which occupies practically the same position. Although the wasps come freely and in numbers to these flowers, the plant does not trust them too implicitly to do its work. When fertilisation is effected, the stigma withers, but until then it remains stretched out over the lower lip, and should no insect do what is needed before the anthers are mature these take up a position over the stigma, and when their pollen is shed some of it is sure to fall upon the stigma. To leave no doubt that the black scale in the roof of the flower is the retrograded fifth stamen, it sometimes develops a more or less—usually less—perfect anther producing a little pollen, but it is rarely that it opens and discharges this pollen. It is an interesting fact that wasps instead of beginning at the lowest flower on a spike, as we saw the bee doing in the case of the Foxglove, commence with the highest one, and this explains why wasp-flowers should mature their stigmas before the anthers. In the last flowers visited on one plant the wasp gets laden with pollen, with which it fertilises the newly-opened blossoms of the next plant visited.

The Water Figwort (*S. aquatica*) is another species common in ditches, and on the margins of ponds. It is a larger, much taller growing species, but otherwise similar in many respects, including the arrangements of its flower-parts and its insect-visitors.

The Yellow Monkey-flower (*Mimulus luteus*), beloved of window-gardeners, though found growing apparently wild by riversides in many parts of the country, is not a native plant, but an introduction from North America in 1826, whose minute seeds would appear to be adapted for water-carriage. We will merely note in passing that the two plates into which the stigma is divided are sensitive, and close together when touched. This appears to be a precaution against pollen falling off when once deposited thereon, for the plates immediately close over it. But there is no doubt another reason for it, and that is that after fertilisation the stigma shall not at all impede the bee's access to the anthers, which stand just behind.

The neat and delicately-fashioned Cornish Moneywort (*Sibthorpia europaea*) has very minute flowers, with a five-lobed corolla of pinkish hue, and four stamens. It appears to be fertilised by small flies, and is particular to deter creeping insects by developing hairs on the tiny round leaves and thread-like creeping stems.

In this family there is a group of no less than six native genera distinguished by one common habit— they are parasitic upon the roots of other plants. They further agree in having the corolla divided into an upper and a lower lip, in having but four stamens, and in the stigma being unequally divided into two lobes. All these things point to a common origin for all the species included in these six groups. They have all become adapted for fertilisation by bees. We will briefly pass these genera in review.

The Cow-wheats (*Melampyrum*) afford another example of science embalming an old error in Greek and letting it serve for a scientific name. The plants have no relation whatever to the wheat whereof flour is made, nor is it a favourite food of cattle: rather is the prefix "cow" meant to indicate that the wheat-like seeds produced by the plant are spurious and worthless. The Purple Cow-wheat (*M. arvense*) grows in cornfields, and its seeds are like black wheat-grain. Upon this foundation there grew up a legend that if any of these seeds were threshed out with the wheat and ground into flour, all the bread made from it would be black. Wherefore do serious botanists speak of the plant by the name Melampyrum, which they got from two Greek words, *melas*, black, and *puros*, wheat.

Only one species is widely common, and that is the Common Yellow Cow-wheat (*M. pratense*), with slender leaves, and two-lipped tubular corollas that become relatively wide at the mouth. At the bottom of the tube honey is secreted by a basal expansion of the ovary, and to enable humble-bees with their long tongues to get at this nectar-store the mouth of the corolla is so wide that they can put their heads inside. Looking at the flowers as they grow, this statement would strike one as an error, for clearly the corolla is too slender for a humble-bee's big head, especially when a couple of yellow pouches are seen to occupy part of the lower lip, and so to restrict the entrance. Closer inspection will reveal folds or pleats in the sides of the tube and in the upper lip. When a bee attempts to get at the honey and presses its head forwards for this purpose, the pleats open out and admit the head of the friendly bee. Now, small bees who lack either the strength to do it, or the sense to see that it has to be done, do not patronise the flower, do not even visit the plant.

The Common Humble-bee (*Bombus terrestris*) cannot manage to get the honey in the legitimate way; but being a business-like bee, intent on getting honey, he bites a hole through the corolla just above the calyx, and through this hole his tongue easily reaches the honey. The four stamens rise up and meet in the vaulted upper part of the corolla, and the anthers lock together by means of the hairs by which they are fringed, so that they enclose a central space into which they shed their light, dry pollen. Each anther-lobe is provided with a stiff, pointed appendage which hangs down into the mouth of the flower, and the inner edge of the filaments is beset with sharp teeth. It will thus be seen that what with the yellow pouches of the lower lip, the sharp points on the filaments and the combined anthers, the actual passage left from the mouth of the flower to the honey is very narrow, and unless the bee is to hurt his tongue by contact with the filament-teeth, he must steer a very even course along the centre of the passage. And this brings his proboscis into contact with the stigma which bends down in front of the anthers, the style lying in the ridge of the upper side; and after that has been done with the probability of placing pollen on the stigma from an earlier flower, the bee's proboscis touches against the down-projecting processes from the anther-lobes, separates the lower edges of the lobes, and a little shower of the dry pollen is shaken down upon its proboscis, thus priming it for a similar office in another flower. Yet in spite of this highly specialised structure, from the plant's habit of growing largely in shady places cross-fertilisation is not by any means invariably effected; but should the insects remain away, the Cow-wheat fertilises itself by the curving downwards of the style until the stigma is beneath the pollen-box, which ultimately opens of its own accord, and pollinates the stigma.

The Yellow Rattle (*Rhinanthus crista-galli*) likewise goes in for showering dry pollen upon the heads of its visitors, but adopts some differences in the details of the mechanism by which this is effected. Here again the yellow corolla is divided into an upper and a lower lip, but the tube is much wider, and is surrounded by the globose, bladder-like calyx. The upper lip is helmet-shaped, compressed from the sides, and arched above. Honey is secreted at the base of the ovary, and in order that the long style shall not come in the way of honey-seeking bees, it keeps close to the roof of this vault, and the stigma protrudes from the narrow opening of the flower. The lower lip serves as an alighting-platform, before which is the narrow entrance to the upper lip, and within the filaments of the stamens rise up, their inner edges bearing sharp spines. The anthers come together just below the style, and open on their inner face, the pollen being prevented from falling by a fringe of stiff hairs. A bee thrusting in his proboscis low down, would be repulsed by the spiny filament-processes, and compelled to try higher up, where he first touches the stigma and pollinates it, then pushing between the anthers disarranges them, and brings a shower of pollen in a narrow line above his proboscis.

There are two races or sub-species of the Yellow Rattle, growing side by side, and distinguished as *major* and *minor*. That which we have described is *major*, the normal form, and this is so freely visited by insects and so invariably gets cross-fertilised that it has lost the power of self-fertilisation—the stigma growing out beyond the anthers where their pollen cannot possibly fall upon it. But the corolla of *minor* scarcely exceeds the calyx in length, and it consequently loses so much of its attractiveness that insect-visitors are comparatively rare, in spite of the fact that the shortness of the tube renders the honey accessible to short-tongued bees as well as to those with long proboscides. This small flower evidently failing to become more attractive, gets over the difficulty with a little ingenuity : the style increases in length, and curls inward so that the stigma comes under the anthers, which discharge their pollen upon it, and so self-fertilisation is effected. Here again, in two forms of a plant with similar mechanism, and growing in the same places under like conditions, we have a striking object-lesson of the advantages derived from a slightly increased size in securing cross-fertilisation. This conspicuousness is increased by the tip of the upper lip in *major* being blue, whilst in *minor* it is white. When the flower hangs down

a little, the upper lip is thought to resemble a nose—
of the Hebrew type. The scientific botanists, with a
sense of humour for which the public fails to give
them credit, have seized upon this likeness and called
it Nose-flower in Greek—*rhin*, nose, and *anthos*,
flower. The Swedish peasants declare that when its
seeds rattle in the bladdery calyx they know their
hay is ripe for harvesting.

Of the genus *Bartsia* we have three species, but
the Alpine Bartsia (*B. alpina*) is a rare northern
species, and the Yellow Bartsia (*B. viscosa*), with its
covering of sticky glandular hairs, is found only in
the South of England and Western Ireland. Red
Bartsia (*B. odontites*) is of far more general occurrence,
and may be found in fields and on the wastes by
roadsides. The deep-pink flowers are borne in pairs
on the spike, and the lower lip serves for alighting
purposes. At the base of this lower lip there are
several purple streaks serving as guides to the honey,
which is secreted at the bottom of the corolla-tube.
The honey is not quite so accessible as it appears, for
though it is but a short distance from the base of the
long upper lip to the honey, yet insects standing on
the lower lip are prohibited from putting their
tongues in at that point, the way being almost
blocked by the broad filaments, which are here thickly
studded with sharp teeth; but higher up, just under
the anthers, the filaments are quite smooth and wider
apart. The anthers come together, and are connected
by the long hairs at the back of each, whilst they
open in front to discharge the pollen. Bees are the
fertilising agents, and when one pushes his tongue in
here, it touches some of the hairs by which all the

anthers are shaken, and pollen falls on the bee's proboscis. In the earliest condition of the flower the stigma protrudes from the opening bud, and is thus very likely to be cross-fertilised before the pollen is ripe. But when the plant grows in shady places, where the visits of bees are not very frequent, they being sun-loving creatures, the style, though at first longer than the bud, does not grow at the same rate as the corolla, with the result that when the anthers are ripe the stigma is between them, and gets covered with their pollen.

In heathy places and old pastures we may find another of these root-parasites, the Eyebright (*Euphrasia officinalis*), the Euphrasy of which our forefathers made so much as an eye-medicine. One old writer (Culpepper) says: "If the herb was but as much used as it is neglected, it would half spoil the spectacle-maker's trade. . . . Arnoldus de Villa Nova saith, it hath restored sight to them that have been blind a long time before." *That* interest in the little Eyebright has all evaporated, because it does not appear to have had any such virtue, but to have been so regarded on account of its name, probably given to it because of the brightness of its flowers in contrast with the grass and its own dark foliage. But to-day we find interest in it by reason of its doubtful method of getting a living, and the mechanism of its flower-parts. The flowers are somewhat similar to those of Bartsia, but the two-lipped corolla is more dilated towards the mouth; painted white, with hair-streaks of purple on both lips pointing to the entrance where there is a yellow spot to make it more conspicuous, and a similar yellow patch covers the middle lobe of the lower lip. The anthers and stigma occupy the upper part of the entrance, but the stigma is mature before the anthers, and so cross-fertilisation is favoured. The anthers are again two-celled, of which the lower cell ends in a long stiff spur; they are all connected together by the downy coats, and the pollen is dry. The filaments are slender, without forbidding teeth, and stand widely apart, to allow a short-tongued insect to thrust his head well into the flower. But the spurs of the anthers hang down in such manner that the insect's head is certain to push against one, and then the anthers are dislocated, and a shower of pollen descends.

This all refers to what we will call the normal form; but there are other forms with smaller flowers, and in these the anthers are ripe before the stigma. In this form the stigma at first stands behind the anthers, but afterwards the style lengthens and brings the stigma just below the anthers, where it will be touched by an insect's head if such a visitor appears, but failing that, the pollen falling from the anthers will effect self-fertilisation. Such self-fertilisation is not possible in the large-flowered form. Bees and some of the larger flies are the insects performing the necessary operations for this species.

In the Louseworts (*Pedicularis*) the corolla is greatly lengthened flattened from the sides, and the internal arrangements still more complex. The colour is a rosy-pink—an advance upon the yellow and white flowered root-parasites we have already referred to, apart from the manifestly more specialised corolla. The upper side is so straight and tubular that it is scarcely fitting to speak of the corolla as two-lipped. The upper portion is in truth more in the nature of a hood within whose vault the stamens are hidden. In the front border of this hood in the Common Lousewort (*P. sylvatica*) there are two tooth-like lobes, and between them the style protrudes. The filaments keep to the sides away from the opening, and the edges of the corolla above the junction of the lower lip are curled outwards and set with sharp teeth, which indicate to the bee that the proper entrance lies elsewhere.

A remarkable feature is to be noticed in the lop-sided way the lower lip is set on the corolla. A casual observer might be pardoned for supposing he was looking at a deformed flower, yet he would find all the specimens he could gather were equally askew. The object of it appears to be to provide a platform useful to bees of different sizes in enabling them to reach the legitimate entrance—where the edges of the corolla are smooth—and so to bring their heads in contact with the stigma. A bee alighting on the oblique platform has to put his head into the entrance sideways, and this is just the method most suited to effect his purpose—that of reaching the honey. But this pressing-in forces the revolute edges somewhat apart, and so alters the position of the toothed lobes above that they force the lower edges of the anthers slightly apart, and the pollen falls on the bee's head. Bees with very long tongues have merely to push their heads just within, but others have to get farther in. The smaller bees are excluded by reason of the long tube; but some of those which cannot reach the honey legitimately bite holes in the corolla low down, and thus achieve their object. The calyx in this species is bladdery, and four of its five lobes are crisped and leafy. It may be as well in leaving this genus to observe that the name Lousewort is due to the belief of our forefathers that sheep feeding where this plant grew were abnormally liable to annoyance by parasitical insects.

We have already alluded to the fact that these latterly-mentioned genera are partial parasites, attacking the roots of various plants and stealing some of their nutriment; but they are not so far gone in rascality as to impose their entire support upon their victim, as a real parasite does. Their retention of normal green leaves is a certificate to the fact that though they steal the crude sap, they elaborate it in their own leaves, and so do not impose so grievous a burden as parasitism implies. But there is an allied group of plants, the Toothworts (*Lathræa*) and Broomrapes (*Orobanche*), which have turned their backs upon honest industry altogether. Perhaps they commenced like Eyebright and Bartsia and Yellow Rattle by merely tapping roots—they are still attached to the roots of their victims, but instead of being content with the almost raw material of which plants are made, they steal the fully elaborated essence, which needs merely moulding into tissues and organs. The difference between the two degrees of vegetable thieves is an exact parallel to "Peter Pindar's" story of the rival broomsellers. One was enabled to undersell the trade by stealing the stuff of which he made his brooms, but was disgusted to find one day he had been undersold in turn. Chancing to meet his rival, he inquired how it was done, acknowledging that he sometimes stole his raw material.

Ivy-leaved Toadflax.

"Ah!" said the other; "I see you have only half learnt your trade—I always steal mine ready-made!" Eyebright, Bartsia, and Co. may some day go the way of Toothwort and Broomrape, which have given up all pretence to honesty, and no longer put forth leaves or any other green organ. Flowers they must have as a means of perpetuating their kind, and these are brown or purple, like the rest of the plant; but the leaves are merely represented by scales.

One word in the title of this chapter has not yet been mentioned, but as we commenced with the Foxglove we will end with the Toadflax. And first let us refer to a species that is not really British, but which has taken so kindly to old walls and stone dykes throughout the country that it is at least as noticeable as any of the truly indigenous species of Linaria. This is the Ivy-leaved Toadflax (*Linaria cymbalaria*), represented in our plate, a plant that has been introduced from the Continent either as a salad or a greenhouse trailer, and has then effected its escape to the garden walls, where it roots in the crevices, and the tender thread-like stems trail down, whilst the wall is well covered by its comparatively large and somewhat fleshy leaves, which are lobed in a fashion suggestive of the ivy-leaf. From beneath the leaves,

Ivy-leaved Toadflax

by the curving outwards of the down-growing flower-stalks, the bright little bluish-purple flowers peep out to the light. These flowers somewhat resemble the garden snapdragon in structure, with the addition of a short curved hollow spur to which honey, produced by the base of the ovary, flows. This is an economical dodge adopted by some plants to adapt them specially for fertilisation by long-tongued insects without going to the expense of increasing the length of the entire flower. The slender appendix involves little extra material, and its narrowness is an advantage, for the insects can suck up the whole of the nectar, much of which might be wasted on a broader surface—in other words, it need not produce more of this attractive fluid than is actually required. The mouth of the flower is closed by a pair of lips which are pressed tightly together to prevent the entrance of small insects bent on taking honey or pollen, or both, without rendering any services in return. On the upper lip of the flower there are a few fine dark lines pointing to the entrance, whilst between two protuberances on the lower lip there is a channel for the insect's tongue. Within, the arrangement of the organs is similar to that of the Foxglove, except that anthers and stigma ripen simultaneously. Fertilisation is chiefly effected by bees, but of much smaller species than those visiting the Foxglove.

All this time the flower has turned as far as possible away from the wall, but fertilisation having been effected, the curve goes out of the flower-stalk, and the swelling ovary hides beneath the leaves. After a time the ovary has developed into the seed-

Yellow Toadflax.

capsule, the seed-eggs have become full-sized seeds and are ripening. Should the ripe capsule open in its present position, the seeds would fall upon the ground, whence the young plants could not attain to the wall-top. But the results are so remarkable that without taking count of the methods by which they are brought about, one is almost justified in declaring that the cleverness of the plant overcomes what would be a disaster to a species that trails down instead of climbing up. As the capsule ripens, the stalk again curls, but this time *towards the wall*, with which the capsule is brought into contact. The curling-up is continued until the further progress of the capsule is stayed by getting stuck against the roof of a chink. There, in due time, the capsule splits open, and the seeds fall into the chink. If these seeds are examined with a lens, they will be found to be covered with ridges and wrinkles, so that they are not likely to roll out of the crevice where, at the proper time, they germinate and provide new material for the beautifying of the unlovely wall.

The Yellow Toadflax (*L. vulgaris*) is an upright-growing plant, with long slender and closely-beset leaves of a glaucous hue, and bright-yellow spurred flowers in a terminal raceme. Before the flowers appear, this plant looks not unlike a densely-leaved Flax-plant, and from this appearance the popular name is derived, the prefix "toad" being given from a fancied resemblance between the mouth of the flower and that of a toad; the name *Linaria* is also suggested by *Linum*, the scientific name of Flax. The orange palate of the lower lip

Yellow Toadflax

is a sign to the bee that the honey lies in that direction. Humble-bees push their heads far in to reach the honey, and in so doing get their backs covered with pollen; honey-bees creep almost entirely in, though sometimes these get the honey in a more expeditious though less honourable way—by biting a hole in the spur.

BUTTER-WORT

IT may at first sight appear to be a singular coincidence that such of our plants as have taken to destroying insects and digesting them, though belonging to widely separated families, agree in being the *habitués* of bogs and marshes. I have already described the carnivorous Sundews as being found in such places, and now have to mention the Butterwort and the Bladderwort, which grow in similar situations. The explanation of the insectivorous habit being common to several dissimilar plants growing under similar conditions is this: the soil in which their roots are found is very deficient in nitrogenous substances, consequently the roots themselves are very small, and used chiefly for the absorption of water. It is therefore necessary that the plant should obtain its nitrogen in other ways, and it has developed the necessary mechanism for catching, killing, and digesting insects, the animal matter thus obtained being afterwards absorbed and assimilated by the plant.

Butterwort (*Pinguicula vulgaris*) is a small perennial growing in bogs, the succulent oblong leaves forming a rosette close pressed to the ground, from amid which the one-flowered scapes rise to a height of five or six inches. The violet flower is two-lipped, with a slender spur. There are only two perfect stamens, and these are attached to the base of the corolla-tube. The style is short and thick, with an unequally two-lipped stigma. The most interesting feature of the plant is its leaves, which are about $1\frac{1}{2}$ inch long, and covered on the upper side with glandular hairs which secrete a sticky, colourless fluid, capable of being drawn out into threads a foot and a half long after the leaf has been excited. This secretion has long been noticed, and the names Butterwort and Pinguicula (from *pinguis*, grease) have been suggested by it. But the ancients knew nothing of its importance to the plant, which indeed was not suspected until Mr. W. Marshall told the late Charles Darwin that he had noticed the plant as growing in the mountains of Cumberland with many insects adhering to the leaves. This led the great naturalist to investigate the reasons for their presence, and the behaviour of the leaves. He found that dead insects and other nitrogenous substances excite the glands to increased secretion of the viscid fluid, which then becomes acid, and capable of

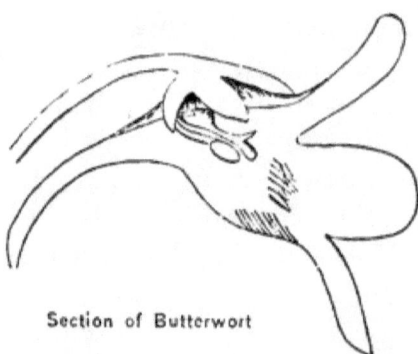

Section of Butterwort

digesting such animal matter as albumen, fibrine, etc. Afterwards, the dissolved nitrogenous matter is absorbed by the glands and assimilated by the plant. The acid nature of the fluid is not exactly a new discovery, for the Laplanders had long used the leaves for the purpose of curdling milk, though they were probably ignorant of the manner in which these acted.

When insects get caught by the stickiness of the leaf, sufficient irritation is set up by their presence to cause the margins to fold over upon them, and the glands of the portions in contact then secrete more copiously than before. There is a natural tendency for the lateral margins of the leaf to curve slightly inwards and form a channel to collect partially-digested insects washed by rain from the centre of the leaf; the "bouillon" resulting from digestion in the more central portion of the leaf also flows towards the marginal channels by whose glands it is absorbed. Meat, skimmed milk, and other animal substances caused the increased activity of the glands, just as they do in the case of the Sundews; but an advance on the Sundew is manifested by Butterwort in its power to digest many seeds, pollen, and some leaves. The seeds that cause no apparent excitement of the glands are those invested by tough coats. There are two other British species.

The flower is adapted for fertilisation by bees, which alight on the lower lip and thrusting their proboscis beneath the upper lip to reach the honey in the spur, first touch one lobe of the stigma with their back, and afterwards get the same part dusted with pollen from the anthers; and to prevent this pollen being detached by the stigma of the same flower, the stigma-lobe is pushed out of the way by the retreat of the insect. In *P. lusitanica* self-fertilisation is effected by the stigma-lobe curling over into the anther-cells and thus getting pollinated. *P. alpina* is fertilised by flies; it has pale-yellow flowers with hairy throat, and it is no unusual thing for large species to get caught and to perish for want of the necessary strength to extricate themselves. It would be interesting to learn by further observation whether this insectivorous plant has learned to make use of insects caught by its flowers as well as those captured by the leaves.

The Bladderworts (*Utricularia*), of which also we have three native species, have yellow flowers similar in structure to those of the Butterworts, with the addition that the stigma is irritable, and its lobes fold up immediately they are touched—a precaution against fertilisation with pollen from the same flower. But the chief point of interest in the Bladderwort lies not in the flowers but in the bladders, from which it derives both popular and scientific names—*utriculus* being the Latin for bladder or bottle. The plants grow in the waters of foul ditches and ponds, where there is an abundance of minute life. They are quite destitute of roots, and float submerged a little below the surface of the water. Each leaf is reduced to a number of delicate threads, on the same principle that governs the submerged leaves of Water Crowfoot (*Ranunculus aquatilis*) and other aquatics.

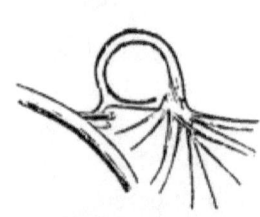
Bladderwort's trap

Near the base of the leaf there are one or more little bladders, bearing a singular likeness to some of the freshwater crustaceans popularly known as Water-fleas—a resemblance largely due to the presence of two bristle-bearing appendages at the end farthest from the foot-stalk. The bladders are translucent and pale-green, and they vary in size, in the different species, from one-twelfth to a quarter of an inch across. Between the appendages—to which Darwin gave the name of antennae—and the foot-stalk on the lower surface there is an opening closed by a valve, and on each side of the opening there are four long bristles. These, together with those borne by the antennae, form a kind of net, whose function appears to be to check the onward progress of small creatures and induce them to pass into the bladder, which they do by pressure against the valve. This valve is a clear, colourless, elastic plate, whose free margin moves inwards in order to open the entrance and admit the low forms of life that are ever pushing in. Its inner face is covered with glandular processes, which appear to absorb but not to secrete. There is apparently no digestive process as in Butterwort, the entrapped victims being killed by lack of oxygen, and retained until decomposition has rendered them capable of being absorbed.

About thirty years ago Mr. Holland remarked to the High Wycombe Natural History Society that "water insects are often found imprisoned in the bladders," and suggested that their bodies were "destined for the plant to feed on." It was at least partially due to this observation that Mr. Darwin was induced to experiment with the plant, to which he was further drawn by its well-known relationship to the Butterworts. Mr. Holland's observation and supposition were fully supported, not only by Darwin, but also by Cohn, who in 1875 published a memoir on the subject. Speaking of the first observations made by him and his son, Darwin says: "My son examined seventeen bladders, including prey of some kind, and eight of these contained entomostracan crustaceans, three larvæ of insects, one being still alive, and six remnants of animals so much decayed that their true nature could not be distinguished. I picked out five bladders which seemed very full, and found in them four, five, eight, and ten crustaceans, and in the fifth a single much elongated larva. In five other bladders, selected from containing remains, but not appearing very full, there were one, two, four, two and five crustaceans. A plant of *Utricularia vulgaris*, which had been kept in almost pure water, was placed by Cohn one evening into water swarming with crustaceans, and by next morning most of the bladders contained these animals entrapped and swimming round and round their prisons. They remained alive for several days; but at last perished, asphyxiated, as I suppose, by the oxygen in the water having been all consumed."

MINT AND THYME

THE family to which belong Mint and Thyme includes such well-known plants as Sage, Woundwort, Balm, Dead-nettle, and Bugle. The flowers of this family (*Labiatæ*) resemble some forms of the Foxglove family — especially the Rattles, Louseworts, and other root-parasites, but from those the Labiates are readily distinguished by the characters of the fruit. In the foregoing family the seeds were small and numerous, contained in a capsule; but in the Labiates there is no capsule, the fruit consisting of four or less nutlets, each containing a single seed. Another difference in the two families is found in the properties of the plants. Those of the Foxglove family, though often very ornamental, include very few of a useful character and some that are highly poisonous, whereas the Labiates are for the most part of use as kitchen herbs, or on account of their aromatic properties, due to the presence of essential oils in glands of the leaf. Other distinguishing characters will be found in the square stems and the opposite leaves.

We have no less than seven species with several sub-species of Mint (*Mentha*) in this country, and these, from their variability, are not always identified without difficulty. They are all distinguished by their strong odours, which have the effect of protecting them from being eaten by most plant-feeders, to whom the aroma is objectionable. This in truth is the reason why these plants have taken to develop the essential oils which the perfumer finds so valuable. The flowers in all the Mints are almost regular, somewhat bell-shaped and small, four-lobed, though the calyx is five-toothed. The upper lobes are broader than the lower, and it is probable that this marks a falling off or degeneration from a condition when the flower was larger and distinctly two-lipped, like its related Labiates. There are four stamens, distant from each other, and the style is not central.

One reason for supposing they have degenerated may be found in the fact that they produce few healthy seeds, the rootstock having taken to branch out and run freely beneath the surface. This habit of increasing by *stolons* may have led to the decrease in the size and showiness of the flowers, the plant having become to a certain extent independent of them for perpetuation. With the exception of the Corn-mint (*M. arvensis*), they are all plants of wet places, such as the margins of ponds, riversides, and marshes. This exceptional species may be found in cultivated ground and drier wastes, and bears two forms of flowers—small ones, in which only the style is developed, and larger ones, in which at first the style is immature, whilst the anthers are ripe: but after these have shed their pollen, the style develops, and the two stigmatic lobes mature and separate. Honey is secreted by the base of the ovary, and is protected from rain by a number of hairs extending from the walls of the corolla. Short-tongued flies are the chief fertilising agents. As might be expected, the larger flowers are visited first.

The very common Water-mint (*M. aquatica*) has a similar arrangement of the floral organs, but instead of the large and small flowers occurring in equal numbers, as they do in the Corn-mint, the larger ones are very much more numerous in the Water-mint. The honey, too, is less accessible to short-tongued insects, on account of the corolla-tube being longer; but as the flower-spikes are denser and taller, more insects visit them, including several of the smaller bees (*Halictus*).

Gipsy-wort (*Lycopus europæus*) is another marsh-loving plant, with minute whorled flowers producing honey, and maturing their anthers earlier than the stigmas. The bluish corolla is spotted with purple in the way of guides to the honey, which is protected by hairs. This also appears to be a degenerate flower, for owing to its small size it has taken to develop only two of its stamens, which are placed on opposite sides, as far away from the style as possible, so that there is little likelihood of self-fertilisation taking place—even if the anthers and stigmas matured together, which is not the case, for the stamens are withered before the stigmas mature.

Marjoram (*Origanum vulgare*), which grows in tall masses on dry banks and downs, has larger flowers, with a very distinct division of the corolla into an upper and a lower lip. When the purple flower opens, the four stamens are ripe, and extend considerably beyond its mouth, whilst the style is short, and its two stigma-lobes are pressed together and immature. The plentiful honey is accessible to many insects owing to the open character of the corolla, and the organs being extended beyond the mouth of the flowers, and these densely gathered into cymes, the carriage of pollen by the insects that crawl from flower to flower is insured. After the pollen is shed, the style lengthens until it is larger than the stamens, and then the stigma-lobes spread apart and equally come in the way of insects. Here again we find smaller flowers with the stamens in an aborted condition; but in this case the large and the small flowers are borne by different plants.

A precisely similar state of affairs is discovered in the more humble Wild Thyme (*Thymus serpyllum*), whose rosy-purple blossoms have a compensation for their smaller size in their sweeter fragrance and scented honey, which causes them to be more readily found and their locality remembered by honey-seeking insects. Just as there are female flowers (those in which the stamens are aborted), there appears to be a tendency to produce male flowers, for in some of the larger form the style never matures. Wild Thyme has completely lost the power of self-fertilisation. In spite of its extremely low stature it is a shrub.

We have two native species of Sage (*Salvia*), of which the larger and more interesting is unfortunately rare. This is the Meadow Sage (*S. pratensis*), which bears its bright-blue flowers horizontally in whorls. The lips are greatly developed, and with special reference to the visits of humble-bees. The lower lip forms a convenient alighting-platform, and the helmet-shaped upper one arches over it and accommodates the stamens and pistil in turn. Only two of the stamens are developed, the others being discoverable as mere points near the base of the perfect ones, which need to be

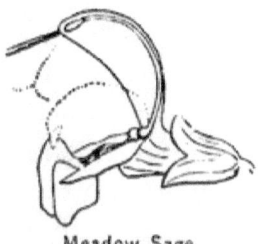
Meadow Sage

described more particularly. Hitherto we have spoken of stamens as consisting of anther-lobes and a filament; but the portion connecting the anther-lobes together and with the filament is known as the *connective*. In Meadow Sage the filament is very short, and the connective so greatly developed that the two anther-lobes stand nearly an inch apart, the upper one occupying the tip of the upper lip, and the lower one (which contains little or no pollen) blocks the entrance to the corolla-tube and the way to the honey. This connective is hinged near its lower end to the filament, and is so beautifully balanced that when a bee pushes its head against the lower anther-

Stamen of Meadow Sage

lobes in order to reach the honey, they yield to the pressure, and in consequence the upper lobes descend upon the bee's back, and leave a patch of pollen there. On the withdrawal of the bee, the lobes resume their former positions, in readiness for further visitors.

At this time the style, which is immature, and also lies in the vault of the upper lip, stands almost straight out from the tip of the lip; but when the pollen has all been distributed, the style lengthens and curves down considerably, the long stigma-lobes curving widely apart, and occupying such a position that upon a bee now visiting the flower its back sweeps the stigmas, and if it has previously visited a younger flower cross-fertilisation is insured. Sprengel described this wonderful mechanism more than a hundred years ago, and gave a figure of the flower with a bee creeping into it. For nearly seventy years Sprengel's book was neglected, and most naturalists were ignorant of his observations and discoveries. One can sympathise with the Rev. Prof. Henslow in this connection, for many of us have had somewhat similar experiences: from his examination of the flowers of Sage he formed a correct estimate of the significance of the mechanism, and wrote to Mr. Darwin describing his discovery, which he imagined to be quite new; but the great naturalist, who was well read in all that his predecessors had done, kindly referred Mr. Henslow to Sprengel's figure and description.

One other species is the Clary (*S. verbenaca*), with smaller purplish flowers, a shorter upper lip, and a habit of self-fertilisation. This is clearly another case of degeneration from a more specialised form of flower, for the connective is still enlarged, though the lower anther-lobe is not developed at all, and the stigmas simply curl towards the anthers and get self-fertilised. Here, then, is another illustration of the point to which we have previously called attention—that the species showing manifest degeneration from a much more highly specialised form have learnt that self-fertilisation means a more extensive progeny and greater colonising power, though accompanied by small stature and inconspicuous flowers. Clary is common, and well distributed over these islands, whilst the more imposing Meadow Sage is rare, and confined to three counties (Cornwall, Kent, Oxford).

It is really amusing to observe the methods by which flowers seek to check the depredations of insects that cannot perform the needed fertilisation, but yet have a decided taste for honey or pollen. Sometimes these insects prove too clever for the plant, and though their entrance may be barred, they contrive to get at the honey by burglary—a breaking through the walls. I have already given instances of this in the case of short-tongued bees, who bore holes in the corolla-tubes and get at the honey. Similar evasions of the plant's regulations have been noticed in *Salvia*. The common white butterfly stands at the entrance to the flower, and insinuating its long slender proboscis beneath the lower anther-lobes, reaches the honey without disturbing the machinery; but this may some day cause the flower to develop some provision, especially against the dishonest butterfly.

The familiar Ground Ivy (*Nepeta glechoma*) has flowers of similar structure, and like the Sage has them of two forms—larger and complete, smaller with perfect female organs only. From the shortness of the tube in the smaller flowers all our native humble-bees can reach the honey with their tongues. In the larger form the mouth of the tube is dilated to admit the bee's head, and make up for the increased length of the tube; but this has the effect of shutting out the Ground Humble-bee (*Bombus terrestris*), whose tongue is short. Cross-fertilisation is insured by the earlier maturing of the anthers.

Self-heal (*Brunella vulgaris*) has dimorphic flowers also, but exhibits a peculiar difference in the form of the stamen: instead of ending in two ordinary anther-lobes, it divides at the tip into two short branches, on one of which are the anther-lobes, whilst the other branch ends in a point which presses against the concave surface of the upper lip, and keeps the anthers in the proper position for pollinating an insect-visitor.

Woundwort (*Stachys sylvatica*) has the upper lip of the corolla small, but as the flowers hang almost horizontally round the stem, it is sufficiently large to protect the organs; whilst the large lower lip forms an alighting-platform. The anthers discharge their pollen whilst the style is only half its ultimate length, but when this attains its full growth the stigmas are widely separated and hang below the anthers, so that they are in the way of insects bringing pollen from a younger flower. In the absence of such visitors self-fertilisation takes place by the stigmatic branches bending against the anthers where a little pollen is still left. Similar but slightly varied arrangements will be found in Marsh Woundwort (*S. palustris*) and Betony (*S. betonica*).

The Hemp Nettle (*Galeopsis tetrahit*) has a vaulted upper lip covering stamens and pistil, and a large lower lip with honey-guides consisting of a yellow spot and a network of red lines. At the base of the lower lip there is a pouch on each side which confines the entrance to the tube to the width and shape of the bee's head. The pistil lies between the anthers, which have already burst when the flower opens, though the stigma is not ready. The latter consists of two lobes, of which the lower and longer hangs down at a later period and touches the bee's back at the spot where it has received pollen from a younger flower. The other British species agree generally in their arrangements.

We have five native Dead Nettles (*Lamium*), which have no relationship with the Stinging Nettles (*Urtica*), and have only been called Nettles because of a superficial resemblance between the leaves of the two genera. The most striking form is the White Deadnettle (*L. album*), which is specially adapted for fertilisation by large bees, notably humble-bees. The outline of the corolla is peculiar, the back wanting in the continually graceful curve that one feels it ought to have. It is a bit flat here, and there is an abrupt turn there, which give the impression of partial distortion; but on careful examination it will be found that these departures from the smooth flow of the curve really help in the special adaptation, whilst they put hindrances in the way of short-tongued insects reaching the honey which they are unable to earn by fertilising the seed-eggs. The upper lip is helmet-shaped, and holds the stamens and pistil as well as fitting the bee's back; the lower lip forming the platform as usual, whilst a pair of side lobes keeps the insect's head in the right position. Honey

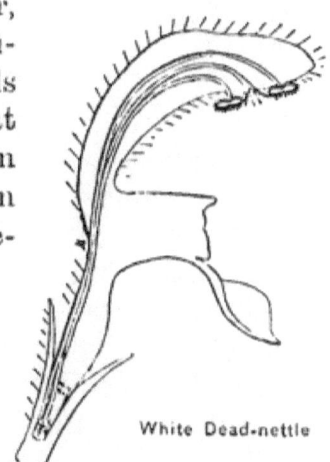

White Dead-nettle

is produced by the base of the ovary, and retained in the lower part of the long corolla-tube. Both anthers and stigma mature together, but one of the lobes of the stigma hangs down between and below the anthers so that it comes first into contact with the back of a bee. The flowers are great favourites with bees, so that cross-fertilisation is no doubt the rule; but those with tongues less than three-eighths of an inch long cannot reach the honey.

The plant frequently grows among Stinging Nettles or in similar situations, and there can be little doubt that natural selection has produced the likeness to *Urtica*, by preserving the plants that most nearly resembled the Nettle from those mammals and insects to whom the stinging powers of Urtica are objectionable. To human eyes the difference in the two species is very marked when the flowers are present, but when these are absent it would take a very sharp-eyed botanist to detect a Lamium growing in a clump of Urtica unless he were at very close range. Small insects could easily creep down the tube and steal the honey, but a ring of hairs above the honey prevents a successful termination to the enterprise; the Ground Humble-bee, however, whose tongue is too short to reach the nectar legitimately, bites a hole through the tube, and other short-tongued species make use of the same opening.

The Yellow Archangel (*L. galeobdolon*) is, like White Dead-nettle, a perennial species, growing more in the copse and hedgerow. Its arrangements are broadly similar to those of its white relation, but the upper lip is so narrowed to the base that it appears to be stalked.

The rosy-flowered Henbit Dead-nettle (*L. amplexicaule*) has a long slender tube, and its proper flowers appear from April to August; but earlier in the spring, and again in autumn, it produces flowers that never open (*cleistogamous*). In these the style is doubled on itself because of the confined space, and the long stigma-lobes twist about among the anthers, getting fertilised by their pollen. The normal flowers also appear to favour self-fertilisation, but it is probable that they are often crossed, though insects cannot be seen around it in any numbers. The Red Dead-nettle (*L. purpureum*) has smaller flowers, in which the anthers and stigmas mature together, and both cross- and self- fertilisation ensues. The tube is shorter, so that bees with tongues of less length than those mentioned above can obtain the honey.

Black Horehound (*Ballota nigra*) protects itself from herbivorous animals by an unpleasant odour calculated to produce feelings of nausea in any that bruise its stem or leaves. In truth, the old Greek name Ballota is said to be derived from *ballo*, to reject, because all animals refuse it as food. The upper lip of its tubular red-purple flowers is hairy within and without, and the lower one is marked with white honey-guide lines. The flower is visited by numerous insects in spite of the offensive smell; these are chiefly bees and butterflies, and they effect cross-fertilisation. Various species of flies also visit it, attracted doubtless by the vile odour, but anxious to get their fill of the honey before they leave—a treat denied them by a ring of hairs set around the tube in such wise that they cannot pass the expanded tip of their tongues through the circle. The pollen is discharged somewhat in advance of the stigma's maturing, and in fine weather is rapidly cleared off by bees before the stigma-lobes separate, so that cross-fertilisation is secured; but in wet weather, with the consequent absence of visitors, the pollen in falling out of the anthers is caught by the hairy lining of the upper lip, and should insects continue to keep away, the stigma twists back and gets pollinated from this source.

The Wood Sage (*Teucrium scorodonia*), with its sage-like leaves and greenish-yellow flowers, is worth a little notice, though so unattractive in the hedge

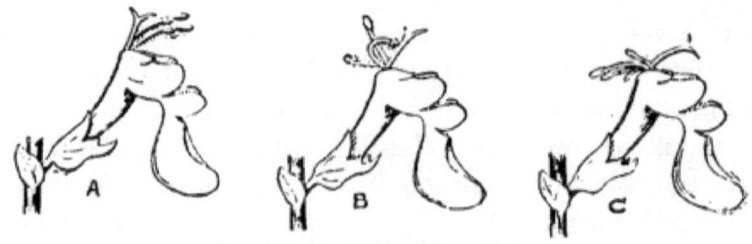

Three stages of Wood Sage flower

and stony dyke. The upper lip looks as though it had been cut off squarely near its base, leaving a wide open tube, above which extend the stamens and pistil. When the flower opens (A), the anthers shed their pollen, but the stigma is not yet mature, though its lobes are already separate. At this stage the stamens stand forward within the flower so that they come into contact with the bees that chiefly perform the work of cross-fertilisation for the flower.

Until the pollen is all shed the stamens maintain this position, whilst the pistil leans over the edge of the flower at the back, so that it is just as far away from the anthers as it can be, and there is little danger of self-fertilisation. When the pollen has all departed (c), stamens and pistil change places: the stamens lean over the back wall and curve downwards, so that there shall be no chance even of the last remaining pollen-grain being transferred to the stigmas, which have now leaned forward over the corolla-tube. The greenish-white colour of the flower, coupled with the fact that it is a favourite with bees, would appear to oppose the view previously expressed in these pages that inconspicuous flowers are as a rule self- or wind-fertilised, but in truth the association of the flowers in this case gives the necessary amount of advertisement, the racemes being quite conspicuous in contrast with the dark foliage and the general surroundings. The abundance of honey also contributes largely to their popularity.

Our last example of this family is the familiar Bugle (*Ajuga reptans*), whose spikes of bright-blue flowers are a conspicuous item in the flora of copse and pasture in spring and early summer. The plant is of lowly growth, which takes possession of the ground by sending out runners from the rootstock. The flower-spike is rendered more striking by the leafy bracts between the upper flowers being usually of a purple hue. The corolla has no more upper lip than the Wood Sage, so that the protruding stamens have to be protected by the bract above. The lower lip is very broad and conspicuous. Anthers and stigmas appear to mature simultaneously, but self-fertilisation is prevented at first by the stigma hiding behind the shorter pair of stamens which protect it from contact with bee-visitors; whilst the anthers all turn down to get in the way of the bees. Afterwards, the lower stamens part sufficiently to allow the stigma to pass between them and come in the way of visitors. Should there be a scarcity of such visits, then the pollen, of course, remains on the anthers, and in all probability some of it will get upon the stigmas from the shorter stamens.

The Ground Pine (*A. chamæpitys*) is a plant of local occurrence in the chalk districts. It has solitary yellow flowers, the lower lip spotted with red, and the leaves are split up into three slender lobes, which have suggested the plant's name from their resemblance to pine leaves. This great departure from the shape of the leaf in the other species is remarkable, but may be explained by an observation of Sir John Lubbock's: he found it growing in the Riviera among Cypress Spurge (*Euphorbia cyparissias*), a plant with leaves similar to the leaf-segments of Ground Pine, and with yellow flowers. Cypress Spurge, like all its family, is rich in a poisonous milky juice, and if it should prove that Ground Pine commonly grows in such company it is easy to understand its departure from the characters of its nearest relations: as in the case of the Dead Nettle, the resemblance to a poisonous plant would have a distinct protective value.

SPURGES AND NETTLES

OUR last words in the foregoing chapter had reference to the protective resemblance of an innocuous plant to one of distinctly poisonous character — the Cypress Spurge; and now we will briefly consider the Spurge family. Although represented in our country by a dozen native species of Spurge, the Box, and two Mercuries, only the Box is of any importance, or at all well known to the public. Yet the family in its cosmopolitan character contains among its three thousand species a large number that are of the greatest commercial importance. Teakwood, castor-oil, caoutchouc, tapioca, and boxwood are among the varied products of this very interesting family, most of whose members produce the milky juice that is intensely acrid even where it is not actually poisonous. In appearance, too, these plants are of the most varied character. In South Africa many of them imitate, so to speak, the Cacti of South America, keeping all their substance in a swollen succulent stem, and reducing their leaves to sharp spines.

The species of Spurge are a puzzle to unscientific flower-lovers who may attempt to make out their floral structure. There are no petals and no sepals, and what appears to be the flower is really the inflorescence or cluster of flowers in an umbel. It is well worth while carefully considering this flower group, and one species is at hand all the summer for the purpose. This is the little Sun Spurge (*E. helioscopia*), which grows in shrubbery borders in the garden, as well as in the waste corners of fields. Its lower leaves are more or less red, but the upper leaves and the floral bracts are of a glaucous hue. At the tip of each shoot and branch there is a rosette of bracts, and in this will be found several smaller rosettes in whose centre is the flower-cluster. If this be picked out and looked at through a lens, the thing will be less puzzling. There is an involucre or general envelope enclosing four male flowers and one female flower. Along the margin of this involucre there are four more or less crescent-shaped glands, and these are commonly taken to be petals. Each of the male flowers consists of a solitary stamen, and if this be carefully regarded it will be found that a joint exists about half-way up. This marks the division between the flower-stalk and the filament. The female flower is a comparatively huge globular ovary with very short stigmas, on a long curved stalk which brings it beside the group of male flowers instead of above them. Honey is freely exhibited on the flat disk around the stamens, and short-lipped insects can obtain it with ease. Flies, bees, and beetles share in the work of cross-fertilisation, but this must take place in a rather uncertain kind of way, there appearing to be no special arrangements such as in so many plants give precision to this operation. One of the folk-names of this plant is Wartwort, from its reputation in rustic medicine as a cure for warts, the milky juice being rubbed upon the offensive excrescence.

Another small annual species found in similar situations is the Petty Spurge (*E. peplus*), with smooth red stems. A more ruddy species and of larger growth is the Wood Spurge (*E. amygdaloides*), with perennial hairy stem and large leathery leaves, occurring locally in copses.

There is one species of very restricted occurrence in Britain which is of similar interest to the Cornish and Mediterranean Heaths previously described. This is the Irish Spurge (*E. hiberna*), whose juice is used by salmon-poachers to poison the rivers sufficiently to kill the fish and cause them to float on the surface. It is found rarely in the hedges and copses of North Devon, and in the south and west of Ireland, but elsewhere only in Western France and Northern Spain.

Flowers of Petty Spurge

The three-valved seed-vessel of the Spurges is worthy of consideration: each valve encloses a seed, and when ripe the valves separate from each other and open on their inner surface, then by contraction the valve ejects the seed with force and throws it far away.

All the plants of this family are left pretty much alone by herbivorous creatures on account of their poisonous qualities, even the insects that will attack them being very few in number. The well-known Box (*Buxus sempervirens*) is even more free from insect-attacks than the Spurges, the odour from the living tree being sufficient to keep them off. This, too, is a shrub of very limited range in this country in a truly wild condition, being confined to the chalk hills of Surrey, Kent, Bucks, and Gloucestershire, though it has become naturalised in many other places. Here, as in the Spurges, the whitish flowers are male or female, borne in small crowded spikes, of which the uppermost are females and the lower males. Here it will be seen is sufficient security against self-fertilisation, for the pollen from the male flowers cannot fall upon the stigmas of the higher flowers. In these Box-flowers there are sepals, but no petals. The male flower has but four sepals and four stamens to match, with a rudimentary ovary; but in the female flowers the sepals vary from four to twelve, and there are three spreading styles whose inner surfaces are the stigmas.

The two species of Mercury (*Mercurialis perennis et annua*) have inconspicuous minute green flowers, with three sepals and from eight to twenty stamens. They grow among weeds at the hedge-foot, and are as a rule left severely alone by browsing animals, who will eat down the plants all round and leave the Mercury standing.

The Nettle family consists of plants whose pollen is carried by the wind, in some cases to distinct plants, so that cross-fertilisation may be effected. There is, of course, little of interest in the mechanism of these flowers, which are all of a simple character, small and green. The family, so far as the British representatives are concerned, includes the Elm, Stinging Nettle, Pellitory, and Hop. The Common Elm (*Ulmus campestris*), so frequent in our hedges and woods, is yet no native, though it has been established among us for many centuries. Its alien character may be guessed by the fact that it never produces seeds in this country, where it is propagated by suckers and layers. The Wych Elm (*U. montana*), on the other hand, is regarded as a true native. Both are tall trees, and live for about five centuries. The flowers are very small, but as they are produced in bunches in early spring they are by no means inconspicuous. They consist of a purplish-brown or reddish perianth, bell-shaped, with five or more lobes and a similar number of purple stamens extending far beyond the bell, and giving to the whole cluster a tassel-like appearance. In the centre of the stamens is the ovary, with two awl-shaped styles, spread out to catch the pollen as the wind blows it from the anthers. It will be evident that for the mere pollination of these styles so many anthers are not needed, but the object of this wholesale production of pollen is to afford the chance of some being carried from tree to tree, to effect a true cross, not merely between flowers on different branches but on separate trees. The flowers are succeeded by seed-vessels of a peculiar form, known as samaras: the ovary develops a wing on each side of the solitary seed, and these wings meet above and below, becoming of a thin, dry, parchmenty character. When the seeds are ripe and the winds blow, these samaras are easily detached from their stalks, and go flying upon the breeze to various distances, until they catch in the hedge or the wind drops momentarily. Thus the old tree gets its children distributed at sufficient distance to give them a chance of growing upon a spot unoccupied by other trees.

The Stinging Nettles (*Urtica*) have developed the art of self-defence to a very high degree of perfection. Though the stems are weak and juicy, and the leaves thin, they are both protected at every point by a close armature of stings constructed on the principle of the adder's venomous tooth. The stings are in this case hollow hairs with very fine firm points, capable of piercing most skins, and at their base is an elongated bulb-like poison-sac. When the point enters the human hand, it gets broken off where the hair is hollow, and the pressure of contact causes the acid poison to be forced up from the bag, through the hollow hair and into the wound, to set up that burning irritation which is so annoying to the incautious meddler with Nettles. Our common Great Nettle (*U. dioica*) is still esteemed by many as a table-vegetable, and no doubt other creatures besides man would eagerly browse upon it but for these virulent stinging hairs. One has but to try the effect of these upon his wrist to understand how animals with tender lips would leave them alone, even when hungry, and to appreciate the cuteness of such plants as Nettle-leaved Bellflower and the Dead Nettle in so growing as to present a very passable likeness to these

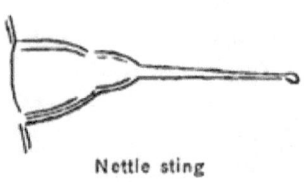

Nettle sting

fiery plants. The Roman Nettle (*U. pilulifera*) is regarded as an alien that has got settled under our walls and on our rubbish-heaps. According to legend, the Roman soldiers who occupied Britain were disgusted with the coldness of our climate, and took to flagellating themselves and each other with Nettles to get up a circulation of their sluggish blood; but our Nettles were considered too weak for the purpose,

Male flower, 1st state

Male flower, 2nd state

Nettle

so they sent home for the more virulent *pilulifera*, and we are invited to believe that thus the Roman Nettle was introduced to Britain!

The flowers are distinctly male and female. In the Small Nettle (*U. urens*) the two kinds of flowers are found in the same panicle, but in the other species they are kept distinct. There is only one floral envelope (perianth), consisting of four green segments which are concave and equal in the male, flat and unequal in the female. When the male flowers are still unopened buds, the stamens are curled up within: when the bud opens, the stamens unroll with force, the anthers burst, and the pollen

Nettle
Female flower

is scattered in a minute cloud and borne by the wind to the stigmas of neighbouring plants. Müller is of opinion that the Small Nettle owes its abundance and wide distribution to three things: "The early period of the year at which it flowers, its regular cross-fertilisation, and the quick ripening of its fruit."

The Hop (*Humulus lupulus*) has interest for us in the fact that it differs so widely in habit from the other members of the family in being a climbing plant. No doubt the founder of the species was a short, erect-growing plant like the modern Stinging Nettle, for all Hops commence life with straight untwisted stems and so remain until they have got their second or third pairs of leaves; then the next joint begins to describe a circle in the air, the free end stretching out and travelling round its base with the sun. But it moves more rapidly than the sun, for the complete circle is described in about one hundred and twenty-eight minutes. This revolving of the youngest joint of the growing stem is at first of an exploratory character: it is seeking a support. Having touched against some upright stick or stem, it begins to curl round it in a spiral direction, and so continues, the rough Hop-stem also twisting on its own axis. This close spiral coil it will be seen is a necessary consequence of the revolving movement of the younger portion of a stem, for in its revolution it meets with an obstacle which arrests the lower part of the shoot, and only permits the upper and constantly lengthening portion to revolve. As the support is a fixture, the revolving shoot is perpetually meeting with resistance. Darwin has given an admirable illustration which makes this proceeding of the tender shoot clear; he says, "If a man swings a rope round his head, and the end hits a stick, it will coil round the stick according to the direction of the swinging movement; so it is with a twining plant, a line of growth travelling round the free part of the shoot, causing it to bend towards the opposite side, and this replaces the momentum of the free end of the rope."

The Hop is fertilised by the wind, and the male flowers are borne by one plant, the female by another. They are all small, and the males, which are produced in long drooping panicles, consist of five sepals with five stamens attached to them. The females form a dense short spike of broad bracts, each bract containing a couple of flowers in its axil, and each flower consisting of a single sepal with an ovary and two awl-shaped purple stigmas which protrude between the bracts to catch the wind-borne pollen from other plants. It will be seen in the female that, as is usual in wind-fertilised (*anemophilous*) flowers, the stigmas are long and spreading as contrasted with the small knob or lobe so common in entomophilous plants, where the pollen is placed with precision by insects.

Sallow Bloom

Beyond the Nettle family lie the important families of Forest Trees: the CUPULIFERÆ, including Birch, Alder, Oak, Beech, Hazel, and Hornbeam; the SALICINÆ, including Poplars, Willows, and Sallow; and the CONIFERÆ, including Pine, Juniper, and Yew. It is impossible within the limits of the present volume to deal with these, but it may be briefly stated that they are nearly all wind-fertilised, with more or less greenish or yellowish flowers.

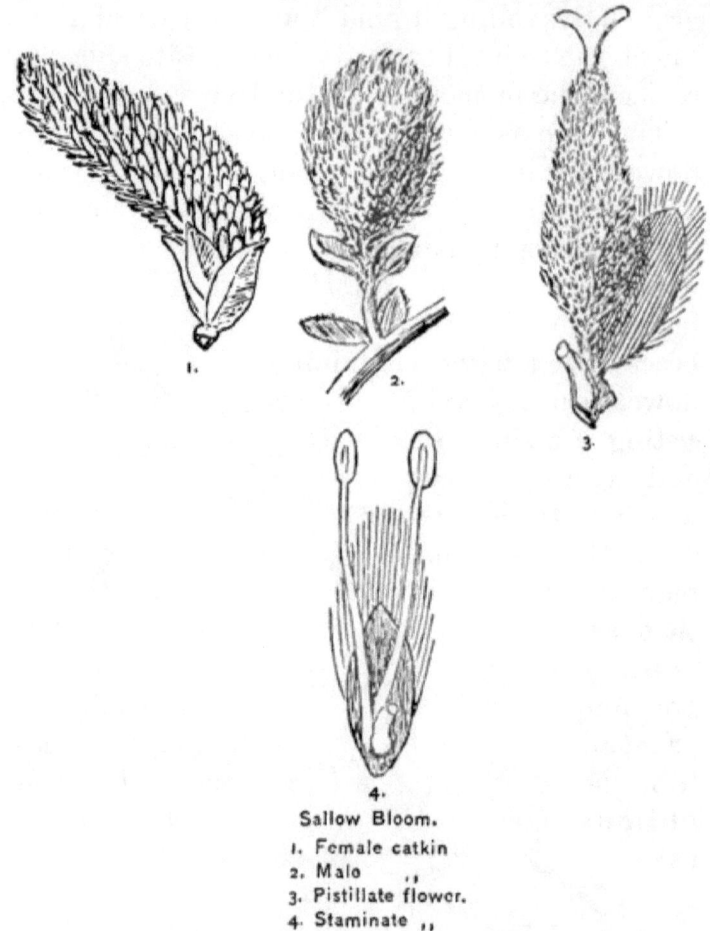

Sallow Bloom.
1. Female catkin
2. Male ,,
3. Pistillate flower.
4. Staminate ,,

Several of the Willows, especially Sallow (*Salix caprœa*), are fertilised by bees and moths. Some of the others are visited by bees for the sake of the pollen.

ORCHIDS

ALTHOUGH everybody has some acquaintance with a few of the beautiful forms among exotic Orchids, and knows that their cultivation has become an expensive hobby, comparatively few know how extensive a representation of the family our native flora affords. No less than sixteen genera with thirty-seven distinct species are found in our woods and fields, and every one of these has an interest of its own. All our native Orchids are terrestrial— we have none of the species that grow on trees and develop false bulbs; all ours grow in the soil, and have bunched or tuberous roots, some of them rich in starch. Their flowers are so different from those of other families that some general description is necessary. There is a perianth of six segments, all coloured, of which the three outer ones are the sepals, and the three inner petals. Two of the petals bound the sides of the flower as it grows, whilst the third is different from them in size and shape, and known as the lip. It is usually larger, and is often continued back as a hollow spur, in which no honey is secreted, but whose inner walls are sucked by insects for the juice in their cells. This lip is, properly speaking, the uppermost petal, but by the twisting of the foot-stalk it becomes the lowest. There is only one perfect stamen, and this is so united with the style that the two combined form what is called the column. Careful examination may reveal the rudiments of two other stamens. At the top of the column is the two-celled anther containing two pear-shaped masses of pollen, known as *pollinia*, the individual grains of pollen being attached to each other by elastic threads, and the whole connected to a foot-stalk (*caudicle*) which ends below in a sticky gland. The ovary is three-sided, usually twisted, and the style often ends in a prominence (*rostellum*) at the foot of the anther, beneath which is a sticky surface formed by the union of the three stigmas. Owing to the difference between Orchids and other flowers, this description may be a little difficult, but if compared with a common form like the Early Purple Orchis (*Orchis mascula*) or the Spotted Orchis (*O. maculata*) it will become lucid.

Early Purple Orchis

We have already seen that some plants in order to secure a successful flower-display "save up" their substance for a year and then put it all into flowers soon after their annual period of growth has commenced. Many of our Orchids come into this category, and their money-bags in which their savings are stored are a couple of fleshy tubers at the base of the stem—in some species round or egg-shaped, in others flattened, and with finger-like processes from them. One of the pair is being used up for the flowers and seeds of this year, whilst the other is being filled with starch elaborated by the leaves for next year. But in some species these are something more than hoards of material, for by the peculiar order in which these are formed, the plant is enabled gradually to move away from the place it formerly held, and thus to get into fresh soil. Many of the bulbous plants change their situations in a similar fashion when the soil they were planted in becomes exhausted for them.

Taking one of the common species mentioned above, let us look at the flower more closely. The lip is broad in front and much narrower behind, affording good foothold for a bee, who on arrival inserts his long tongue in the hollow spur and pushes his head against the rostellum. The slight pressure is sufficient to rupture the membrane by which it is covered, and it splits across, disclosing two round disks connected to the foot-stalk of the pollen-masses, and covered with a sticky fluid, by means of which they adhere to the head of the bee.

Early Purple Orchis
Pol. Pollinia
Ro. Rostellum
St. Stigma

At this time the two pollinia are quite upright, but their weight causes them to bend forward until almost horizontal and appearing like two additional antennæ, at which point they remain fixed. Often a bee, finding that he has become burdened with these pollinia, strives to scrape them off with his fore-feet, but rarely with success: the cement hardens in the course of three or four seconds, and they become immovable by such efforts, though when they have drooped forward they can be partially bitten off by the mandibles. When, however, the bee visits another Orchis flower, the pollinia with the greatest accuracy are pushed against the stigma, which being sticky detaches some of their pollen-grains.

This process may be closely watched by lying down near the flowers and waiting for the bees; but it may also be imitated with greater convenience by gathering a spike of the flowers and pushing the point of a pocket-pencil into the spur. It will be found that the wood near the point will touch against the rostellum, and if now withdrawn the pollinia will be seen attached to this part. The partial drooping can then be witnessed, and when this has ceased, the pencil may be pushed into another flower in order to witness how accurately the pollinia strike the stigma. A bee will visit only about four flowers on each spike, and if neither of these has been previously visited it will take away four pairs of pollinia, but the time spent

Early Purple Orchis: side view

in each flower is so short—about four seconds—
that the pollinia have not drooped sufficiently to
fertilise one of them. By the time the bee has
reached another flower-spike the necessary depression has occurred, and the bee therefore cross-fertilises several flowers in succession. This is, no
doubt, the reason why so large a number of Orchid
seeds are produced, most of the abundant seed-eggs being fertilised.

All these details, though given with particular
reference to the Purple Orchis (*O. mascula*), apply
in all essentials to the Green-winged Orchis (*O.
morio*), the Marsh Orchis (*O. latifolia*), the Dwarf
Orchis (*O. ustulata*), and other members of the
genus. Fertilisation is effected in *O. mascula* by
the visits of several kinds of humble-bees;
O. morio and *O. latifolia* by humble-bees, hive-bees, and some of the smaller bees (*Osmia*, etc.).
O. maculata is visited occasionally by humble-bees, but chiefly by the larger two-winged flies
(*Empis, Volucella, Eristalis*, etc.). On 3rd July,
1898, in woods near Dorking, I observed that a
very large number of the flowers of *O. maculata*
had been visited by small flies that had crept right
into the spur, but had been unable to extricate
themselves again. On many spikes examined, nearly
every flower was thus to all appearance rendered
useless. I regret that as I was a long distance
from home I lacked the opportunity or appliances
for dissecting out the fly, or for observing whether
the ovaries developed. This appears to be a serious
matter for the plant, and if the circumstance is at all
common should lead to some new development whereby

the flower may be enabled to exclude such unwelcome visitors, which lose their lives in a vain effort to suck the interior of the spur.

O. ustulata has the entrance to its spur so reduced in diameter that its fertilisation has to be carried out by butterflies and moths. The Pyramidal Orchis (*O. pyramidalis*) is likewise adapted for such insects, in this case by the rostellum overhanging the mouth of the much-lengthened spur and thus closing it to all but the slender tongues of butterflies and moths, for whom also the pollinia are specially adapted, being joined at their base by an adhesive saddle-shaped band to fit the proboscis instead of the head.

The greenish-flowered Man Orchis (*Aceras anthropophora*) is one of the numerous peculiarly-shaped flowers which naturally suggest likeness to some other creature. The lip has four narrow lobes which stand for arms and legs, while the other petals and the sepals form a hood which passes for the man's head. In this flower there is no spur, the pollinia are joined at their base and contained in one pouch instead of two as in the genus *Orchis*.

The genus *Ophrys* includes three species with flowers remarkable for their likeness to animal forms, which has suggested their popular and scientific names. One of these is the Bee Orchis (*O. apifera*), whose lip is of a purplish-brown, and the sepals coloured pink within. The general appearance is not unlike that of a humble-bee, and many years ago Robert Brown suggested that the likeness had for its object the prevention of bee-visits. This is a view that does not commend itself at the present day; but it must be confessed that no satisfactory explanation for the likeness or the conspicuous colouring has been yet advanced. The colour of the lip might lead one to suppose that flies and wasps were invited to fertilise it, but the organs are so modified that self-fertilisation is a certainty. There is no rostellum, and the pollinia are perched on such long, attenuated footstalks that they fall forwards by the weight of the pollen-masses and dangle against the stigmas.

The Fly Orchis (*O. muscifera*) presents a more striking resemblance to an insect. The lip and petals are bright-brown or dark-purple, the former with a squarish blue patch in the centre, and the latter so slender that they look like antennæ, whilst the fly's eyes are found in a couple of black shining bodies at the base of the lip. The plant is visited by two-winged flies such as love carrion, and those are mostly attracted also by purplish-brown flowers. The fact that they visit this plant is no new thing, for

Fly Orchis

more than two hundred years ago, John Parkinson, the king's herbarist, says, after describing the flower, "The naturall flie seemeth so to be in love with it, that you shall seldome come in the heate of the daie but you shall find one sitting close thereon." The flies are attracted not only by the carrion colour, but also by minute drops of fluid which ooze from the surface of the lip, and probably by the "eyes" of the supposititious fly, which, though perfectly dry, are shining and very like large drops of liquid. These "eyes" are licked by the fly in the belief that they are globules of water, and in doing so its head would touch the base of a pollen-gland and detach it by adhesion. Grass of Parnassus (*Parnassia palustris*) has similar fraudulent imitations of nectar globules in its flower.

The Musk Orchis (*Herminium monorchis*), which grows locally on chalk downs in the more southern portions of England, has a loose slender spike of small green flowers. There is no spur or honey, but in the evening the flowers give off a musky odour. Many small insects belonging to several orders—Hymenoptera, Diptera, Coleoptera, etc.—visit these flowers. Of twenty-seven species observed by Mr. George Darwin to visit the flowers, the largest was not more than one-twentieth of an inch long.

The purple-flowered Fragrant Orchis (*Habenaria conopsea*) has a slender spur, and the rostellum distinctly elongated between the glands of the pollinia. Owing to the fineness of the spur, only moths and butterflies can get their tongues into it. Among the moths fertilising it may be mentioned

Butterfly Orchis

the Silver-Y and the Yellow Underwing, which are attracted by the fragrant odour. The Butterfly Orchis (*H. bifolia*) has whitish flowers, and is fragrant only at night. These two facts imply that night-flying moths are specially welcome to it; but on looking more closely we find that the spur—which varies from half an inch to an inch and a half long—is so slender that bees can make no use of it.

The rare Bog Orchis (*Malaxis paludosa*) has tiny yellow-green flowers which are not twisted on their stalk, so that the lip comes to the upper side of the flower, its base incurved and embracing the column. The anther is hinged to the top of the column, and contains four pollinia, which are all attached to one gland. The plant is peculiar in that the leaves produce bulbils at their edges, and these develop into new plants. Closely allied to Malaxis we have two species of Orchids representing two separate genera which are of peculiar interest because they introduce us to a condition and habit of life differing widely from that of the plants we have been considering hitherto. Coral-root (*Corallorhiza innata*) and Bird's-nest (*Neottia nidus-avis*) are known to botanists as *saprophytes*. In order to obtain the carbon wherewith so much of the plant is built up, the carbon-dioxide (or carbonic-acid gas) of the atmosphere is broken up by the activity of the chlorophyll, or green colouring matter. Some plants, however, do not possess this chlorophyll, and these are entirely lacking in the green colour. When chlorophyll is absent from a flowering plant, it is a sign that the plant has given up the manufacture of its own building materials and has learned to obtain them by an easier process. Such a plant may be a true parasite, getting these valuable materials in the raw state by tapping the vessels of its host, or it may be a saprophyte, which feeds upon the decomposing forms of other plants which have already produced these desirable substances by means of their chlorophyll. To such plants roots and leaves would be useless because they would have no functions to discharge, the whole surface being absorbent. We therefore find that, instead of roots, these are merely branching extensions of the base of the stem securing the plant in the soil, and the leaves have been reduced to mere scales of a brownish hue. The Coral-root is so called because its underground branches are fleshy fibres resembling an interlacing mass of white coral; whilst those of the Bird's-nest are so interwoven that they present the appearance of a bird's nest. A consideration of these plants leads to the belief that their ancestors were once ordinary plants with green leaves containing chlorophyll, but that one member of the family found it advantageous to get his carbonaceous material from the lifeless bodies of other plants which had accumulated it in the usual way. Succeeding generations following in the new method of obtaining food at first perhaps partially, then wholly, the leaves —rendered brown by the non-development of useless chlorophyll — would gradually dwindle in size for want of using until they became the scales we now find sheathing the stem. Organs once possessed by a species are seldom given up entirely, however useless they may have become; so that should circumstances again necessitate a return to the older method of breaking up carbon-dioxide, these brown scales may again develop into large leaves containing chlorophyll, and perform the normal functions of leaves.

The Tway-blade (*Listera ovata*), whose twin broad leaves are so noticeable in woods and pastures in spring, presents a strong contrast with its relations just mentioned. Here all is green, even to the long slender raceme of small flowers, and chlorophyll is abundant. The flowers are in most respects similar to those of the Bird's-nest, which, consequently, we have not described separately. The lip is long and narrow, slightly hollowed at the base, and honey is secreted in a groove along the centre. The anther is hinged to the back of the column which is free from the sepals and petals, and the pollen is in two powdery masses. Ichneumon flies and small beetles are the agents in cross-fertilising this Orchid, and the arrangements for utilising their visits afford an interesting variation upon those hitherto considered. These small insects alight on the divided tip and gradually make their way upward, licking up the nectar in the central groove until they come to the end of it at the base of the lip. Then, lifting up their heads, preparatory to moving off, they strike them against the edge of the thin tongue-like rostellum. This touch is sufficient to irritate the rostellum to the secretion of a minute drop of a wonderful white cement, which touches at once the insect's head and the pollinia, fixing them together and hardening instantly. No matter how many flowers the insect visits in which the pollinia are still present, these are added to its head, so that six or seven pairs of these pollinia may be carried upon the head of a single ichneumon fly or beetle before it visits a flower from which the pollinia have been previously removed. Then, the rostellum being no longer irritable and ceasing to protect the stigma, some of the pollen is now left behind. A number of flowers may thus be cross-fertilised before the insect takes up more pollinia.

Lady's-tresses (*Spiranthes autumnalis*) has somewhat similar flowers, but white and fragrant, the lip channelled at its base, and the tip undivided. Its colour and fragrance appeal to a higher group of insects, and its fertilisation is undertaken by humble-bees in spite of the small size—one-sixth of an inch—of its flowers.

The Helleborines, formerly included in one genus, are now divided into the genera *Epipactis* and *Cephalanthera* on account of differences in their floral structure. They have creeping rootstocks, and usually white or whitish flowers. Instead of the lip being continued backwards into a spur, as we find in Orchids fertilised by long-tongued insects, in the Helleborines it forms a basal basin holding honey, more suited to accommodate the heads of wasps and flies. The broad-leaved Helleborine (*E. latifolia*), which is not uncommon in woods in July, has its greenish-white flowers marked with purple and yellow. The anther is hinged to the top of the column, and the pollen masses are powdery, but the pollen-grains are so glutinous that they cannot fall upon the prominent stigma. A wasp may take his fill of honey without touching any irritable surface or getting pollinia attached to his head, but when he has had sufficient nectar and prepares to go, his head *must* come in contact with them, and he is not allowed to go away without taking them with him to fertilise the next spike. There appears to be no possibility of self-fertilisation occurring, and so impressed by this fact was Darwin that he declared *E. latifolia* must become extinct in any district where wasps ceased to exist.

The White Helleborine (*Cephalanthera pollens*), as indicated, is similar to *Epipactis*, except that there is no rostellum. The petals and sepals have their margins curled inwards, and the lip is constricted in the middle, dividing it into a basin-like basal portion and a terminal yellow lobe, which at first folds upwards and closes the entrance to the flower. Some of the pollen-grains adjacent to the edge of the stigma throw out their shoots and partially self-fertilise the seed-eggs. Darwin was of opinion that the principal object of this is to retain the pollen until insects arrive, to which the grains adhere, and by which the flowers can alone be perfectly fertilised.

The very rare Lady's Slipper (*Cypripedium calceolus*) must be my last example of the British Orchids; and this shows considerable variation from the structure and mechanism of the other species. The sepals and petals are long and slender, and they spread widely apart. The lip is developed into a large bag the edges of whose mouth turn inwards. In all the other species I have mentioned a single anther usually

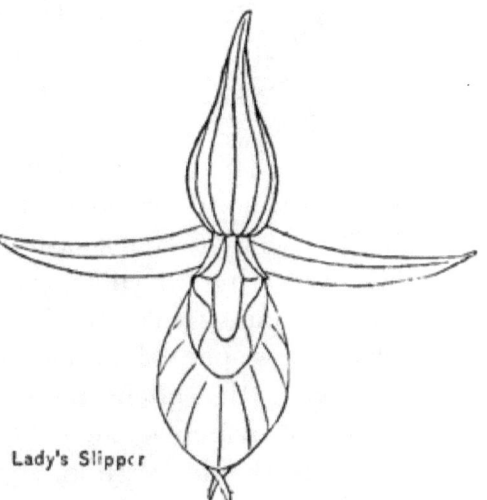

Lady's Slipper

bears two pollen masses. The fundamental plan of all Orchid flowers suggests three stamens, but usually two are aborted and the third developed. Now, in Cypripedium the usually aborted anthers are developed, whilst the third is metamorphosed into a shield-like purple-spotted lobe attached to the extremity of the column and nearly closing the orifice of the pale-yellow lip. The perfect anthers are situated one on each side of this aborted stamen, and the stigma lies below it. These two organs occupy the centre of the entrance to the lip, the opening left being horseshoe-shaped. On the floor of the lip there is a broad band of long hairs which secrete honey. Various species of *Andrena*, a large genus of small pollen-collecting bees, attracted by the colour and perfume of the flower, enter the lip, and after licking and biting the hairs, attempt to leave; but the incurved edges of the entrance present an obstacle rarely surmounted, and the only means of ready exit is at the basal end of the lip, where there is space beneath the anther on each side. The long hairs on the floor enable the small bees to reach these doors, but in getting to them the bee must push its back against the stigma, and in passing through the exit must press one shoulder against the anther, now covered with pollen invested in a viscid fluid, by means of which some of the grains will infallibly attach to the bee's thorax. On seeking escape from another flower this will be deposited on the stigma, and a new supply obtained as the bee leaves by the basal exit.

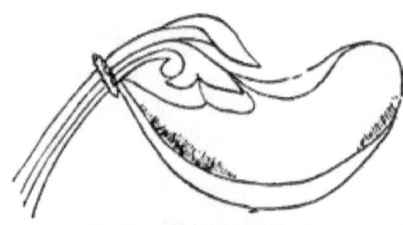

Section of Lady's Slipper

FLAG AND CROCUS

FEW members of this family of flowers honour Britain with their presence as wildlings, and of these only the Yellow Flag (*Iris pseudacorus*) is at all common. They are all perennial plants, and their rootstocks take the forms of tubers, corms, or bulbs, or in some cases are thick creeping rhizomes. The leaves are sword-shaped, usually so folded at the base that a section gives the form of a saddle (*equitant*). The flower consists of a six-parted perianth without distinction between petals and sepals, all being coloured alike. There are only three stamens, and the ovary is three-celled to match, the simple style also ending in three stigmas. The ovary develops into a relatively large leathery capsule splitting into three valves and disclosing the numerous seeds.

For one of the rarest of these plants we must go to the bogs of Galway and Kerry, where in July and August we may see the beautiful little Blue-eyed

Grass (*Sisyrinchium angustifolium*), a plant that has curiously got there from North America, but which appears, in Kerry at least, to be truly wild. Its rooting portion consists of a tuft of rigid fibres, and before the flowers appear the leaves might be mistaken for those of grass. These flowers are less than an inch in diameter, blue inside, and from one to four in an umbel. In America it is abundant in meadows, and its grassy appearance is thus explained.

Romulea (*Romulea columnae*) is similar to the last-mentioned in its rarity as a British plant, but the home of the species is the Mediterranean region, the west of France, and the islands of the Azores, so that its appearance in the Channel Islands and at Dawlish in Devon is not so remarkable as the case of Sisyrinchium. Romulea approaches nearer to the Crocus, and has a corm the size of a pea as an underground base, whence proceed the slender, half-cylindrical, and wiry leaves, and the short flower-scape, with its two or three greenish flowers lined with white and streaked with purple.

Two species of Crocus are naturalised here—the Autumnal Crocus (*Crocus nudiflorus*) and the Vernal Crocus (*C. vernus*), the former considered by some authorities to be a native. As everybody knows, they have underground corms, for these are conspicuous in the seedsmen's shops every autumn, and from these the grass-like leaves rise in a bundle, the outer series wrapping the inner and giving such support that they serve as a stem. In the majority of species the leaves appear with the flowers; such is the case with the Vernal Crocus, but there are others that agree with the Autumnal Crocus, whose flowers appear in the fall of the year whilst the leaves lie dormant within the sheathing bracts until the following spring. This species produces but one flower to each bundle of leaves, but the Vernal Crocus produces several; in both species these are purple, though the Vernal is sometimes white. There is rather a singular point in the colouration of the flowers: in nearly every species of Crocus known the throat of the flower-tube is more or less orange in colour, and this colouration extends only to that part of the interior upon which the pollen falls from the anthers. Mr. Maw suggests that this golden zone may be "an inherited character from the mere mechanical tincture of the fuller orange pollen-grains." This may well be, seeing that the colour lies round the entrance to the honey-tube

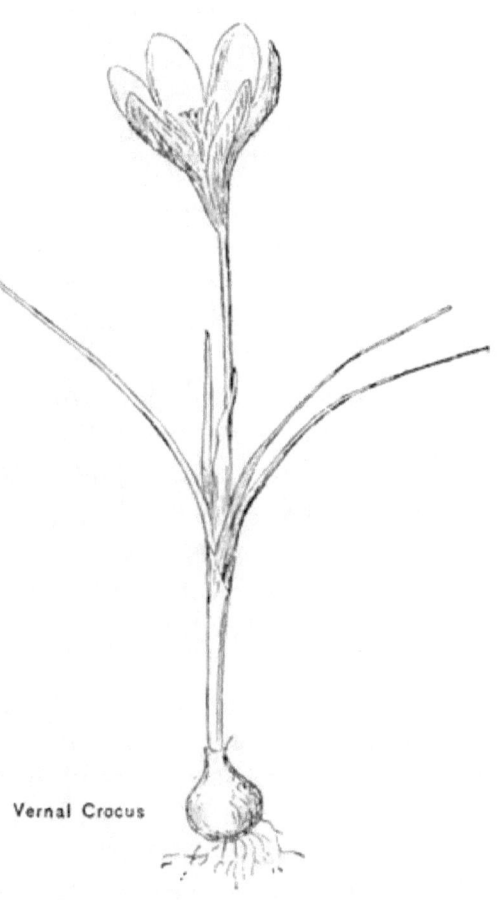

Vernal Crocus

and that such a honey-guide may have proved advantageous to the individuals first exhibiting it. In both our species the anthers and stigmas are orange or yellow. The stamens stand up against the style, the anthers open along their outer faces, and the stigmas arch slightly over them. At first only the anthers are mature, and the moths that alone can reach the honey, using them for alighting purposes, get their undersides dusted with pollen. In a slightly older flower the stigmas by bending apart place themselves in the way of getting this pollen transferred to them; but should the insects fail to appear, the stigmas press down between the anthers in such a way that they dust themselves with pollen. The ovary is hidden away among the bases of the leaves, and not until considerable growth has taken place is it brought up above ground by the lengthening of the flower-stalk.

A peculiar thing is the simultaneous appearance of the pods of the two species: the Autumnal Crocus keeps her pod hidden until the following April, yet the Vernal Crocus that flowers in March has her pod above ground at the same date as the other. All the species of Crocus observe the same rule.

About midsummer, when the seeds are ripe and red, the three valves of the spindle-shaped capsules open and distribute them; but now a difference appears, for the seeds of the Autumnal species germinate in November, whilst those of the Vernal species remain quiescent until the following spring.

There is another way in which the Crocus gets propagated: at the base of each shoot arising from an old corm there will be formed after the flowering period a little corm, so that as each corm produces several leaf-bundles there will be a similar number of new corms clustered on the remains of the old one, and these will, of course, be nearer the surface. Each succeeding year the corms will be more crowded and less deeply buried, until at last they lie close together

Yellow Flag.

on the surface. The time has come for a "redistribution of seats," and as the corms lie dry upon the ground they get scattered abroad by passing animals. But how are they to get buried? The distributed corms having reached a suitable place, send out a thick shoot downwards, and along this shoot the substance of the old corm passes, and forms a new corm below at an appropriate depth.

The stately Yellow Iris (*Iris pseudacorus*), with its clear yellow flowers, is a conspicuous object by marsh and stream in early summer, sometimes covering many acres with its sword-shaped leaves and flower-stems. Peculiar as we found the Orchid flowers, those of the Flag are more puzzling at first sight, and not a little reminiscent of Orchid flowers. To all appearance there are three two-lipped flowers upon one foot-stalk, but a simple dissection shows that these are but three parts of one. All the parts of the flower are joined together at its base, but higher up they are separate, and we can distinguish between sepals and petals, though it is customary to speak of them in this case merely as perianth-divisions. The broad reflexed lower lips are the three sepals, the overarching upper lips are really broadly-winged styles resembling petals — "out-Heroding Herod" in fact, for they are more petal-like than the real petals, which may be found almost erect between the sepals. The stamens, though attached by their base to the sepals, curve in such manner as brings their anthers close to the broad style and just below the transverse shelf-like scale which is the actual stigma. If the lower part of the flower be cut across, it will be evident that the style is really central. Honey is poured forth from the inner surface of the lower parts of the flower, and flows all around the ovary, yet it can be reached only from the sepals, and the insect must have a long tongue to obtain it, although he can push his head some distance between style and sepal.

On examining a number of Flag-plants, it will be seen that there are two types of flower. In one the style lies close upon the sepal, leaving no room for a bee to crawl between; in the other form there is sufficient room for a humble-bee. Now, these two conditions have evidently arisen to adapt the flower for fertilisation by two different insects. Both forms are much visited by the Long-tongued Hover-fly (*Rhingia rostrata*), and the more roomy flowers enable it to obtain honey without fertilising in return; not only so, but before it leaves the flower it reaches up to the anther and feeds on the pollen also. What we may call the bee-form therefore loses by the patronage of *Rhingia* and gains nothing in return. But there are several species of Humble-bees (*Bombus agrorum, B. hortorum*, and *B. derhamellus*) that also frequent these flowers, and these on pushing between style and sepal at first rub their backs against the stigmatic scale and fertilise it with pollen they may have brought from a previously visited Flag; then they rub their hairy backs against the anther, and take up a fresh supply of pollen before they reach the honey. In retreating from the flower they do not retrace their course, but crawl sideways from beneath the style without touching the stigma. In the fly-form of flower Rhingia goes through a process similar to that of the bee in the other form, but when it has done sucking honey it crawls back under the stigma, but has not room to allow it to eat pollen. Only the non-sensitive lower edge of the stigmatic scale is touched as it passes, but no pollen is left upon it; this is retained among the hairs of the fly's thorax and deposited on the stigma of the next flower visited. Of course, there must be a good deal of self-fertilisation effected in this way when the fly or bee carefully visits the three departments of each flower, but often only one is explored, and then cross-fertilisation must ensue; further, where three stigmas are touched in one flower, the first of the trio must be fertilised with pollen brought from the flower last visited.

Our other species of Flag is the Roast-beef Plant, Gladdon, or Fœtid Iris (*I. fœtidissima*), a plant of much less frequent occurrence, in copses on a limestone soil. It is a much smaller plant than the Yellow Flag, with soft dark leaves and purple sepals, the stigmas and petals yellow. It has an evil odour, which has earned several of its names, some persons recognising in it a suggestion of the aroma from a roasting joint of beef, whilst others regard it merely as a "stink."

DAFFODIL AND SNOWDROP

AFFODIL, Snowdrop, and Snowflake are the sole British representatives of the extensive and beautiful Amaryllis family. These plants form a kind of connecting link between the Iris family and the Lilies, though they differ from both. Instead of the solid corm of the Crocus, which is a depressed stem, we have here bulbs, which are stems surrounded by succulent leaves. They serve the same purpose as the corm—that is, they are underground treasuries in which the material made by the leaves this spring may be stored up through summer drought and winter frost to be utilised in the production of rich flowers early next spring. From this base arise the long flat narrow leaves with parallel margins, and the flattened flower-scapes. At the summit of the scape there is an envelope (*the spathe*) of thin parchmenty material containing the flower-bud, or buds which rupture it as they increase in size, when it begins to shrivel. The flower is again a perianth of six portions united below and free above, with six stamens, a three-celled ovary beneath the perianth, and from one to three stigmas.

In the Daffodil, or Lent Lily (*Narcissus pseudo-narcissus*), the solitary pale-yellow perianth is tubular in its lower part, but the free ends of the segments spread, and the mouth of the tube is surmounted by a circular "crown" as long as the perianth-segments, with its edges crisped and toothed. The anthers open towards the style, but from the length of this and the almost horizontal attitude of the flower there is little likelihood of the pollen falling upon the stigma. The flowers appear to be visited by early bees, such as the hive-bee and species of Andrena, in the day-time and by moths in the evening. The colour is adapted for both seasons, for by day the flowers are of gold in the spring sunshine, whilst at evening they look almost white. Moths alighting upon the stigma push between the style and the anthers, and so reach the honey at the base of the tube. Bees alight sometimes upon the stigma, sometimes upon the lower inside of the crown, but in either case they must get dusted with pollen by contact with the anthers, and so cross-fertilise the first flower upon whose stigmas they land.

The Snowdrop (*Galanthus nivalis*) differs considerably in form, though its structure is the same as that of the Daffodil. It has no crown, and the perianth is bell-shaped instead of tubular. The six stamens stand around the style, and as a consequence of the hanging posture of the flower, the anthers open by slits. The tips of the anthers are drawn out into a kind of spur which curves towards the petals, and upon the attempt of an insect to get at

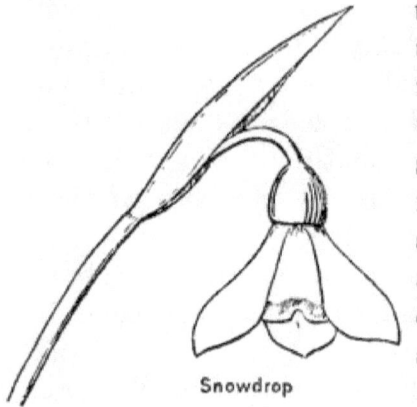
Snowdrop

the honey in the grooves of the petals one or other of these processes must be touched, with the result of shaking out some pollen from the anther. The stigma hangs below the anthers, and from the closeness of these to the style, the pollen would fall upon the stigma if they opened fully along the side. If cross-fertilisation has not taken place through insect-agency, the terminal pores open more widely and let the pollen fall upon the stigma. Hive-bees, who are astir on sunny days even in winter, are attentive to the Snowdrop, alighting on the sepals and crawling thence to the petals, putting their heads over the edge among anthers and stigma, the latter organ being first touched, and the bee's head soon afterwards dusted with pollen wherewith to fertilise other Snowdrops. The hanging position effectually protects the honey from deterioration by rain or snow, and the bees are notified of the good cheer awaiting them by a delicate perfume. Six hours a day are considered by the Snowdrop to be a sufficiently long

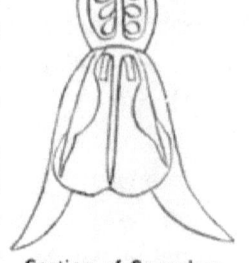
Section of Snowdrop

working day for the bees in winter, so like a Government office it does not open its doors until 10 a.m.,

and closes them again at 4 p.m. It may be noted
that the Snowdrop only puts forth two leaves
and one flower. It has been pointed out to
me more than once as a peculiar thing, that
the Snowdrop flowers keep their freshness
through a long spell of inclement weather,
but soon fade after a day or two of fine
sunny weather. The reason is obvious in the
light of what we have stated: during bad
weather there are no bees about, but in fine
weather fertilisation is soon effected, and after
that the flower fades, being no longer necessary.

Snowdrop's stamen

When mentioning the Lesser Celandine in an early
chapter of this work, I called attention to the importance of the bags of wealth hoarded from one
year to the next by this bright spring flower. The
point there insisted on at some length has also to be
considered now we have reached what are best known
to the general reader as bulbous plants, or more
shortly as "bulbs." Here again the plants are mostly
those that flower quite early in the year, and they
are only enabled to do so by reason of the "much
substance" laid up during the past year. After they
have flowered this year and set seed, the leaves will
lengthen to an extent that is enormous compared
with their size when the flowers appeared. They
have set seriously to work to lay up a reserve for
next year's flowering. Then, when the leaves have
attained to their highest development, their starch
will be poured down into the cells of the bulb, which
will increase the thickness of its layers, whilst the
leaves vanish utterly. Then when the great revival
to activity — the fitly-named Spring — comes, the

plant can rapidly convert its store of starch into cellulose, and produce both leaves and flowers independently even of sunlight. Give them water at their base and surround them with an atmosphere containing oxygen, and they can produce flowers in the dark. Of course, the leaves would have to have opportunity for working in sunlight after that, or the bulb would be useless for another year.

There is one more little group of the Amaryllids represented in this country by the Snowflakes. One of these, the Spring Snowflake (*Leucojum vernum*), is found in Dorsetshire only. It is similar to the Snowdrop, but its leaves are longer and more numerous, its taller scape usually supporting two flowers, the sepals and petals more nearly equal in size, and the sepals having green tips. Another species, the Summer Snowflake (*L. æstivum*), has from two to six flowers, which are produced about May, though its leaves make their appearance during winter. Its flower-buds are erect, but as they open they droop in order to protect the honey and pollen; and when the flower has passed, and the seed-capsules begin to develop, the scape falls prostrate, to take greater care of them.

LILIES & ONIONS

NO doubt at the first blush there does appear to be something incongruous in the mixing of refined and æsthetic Lilies with the vulgar and malodorous Onion and Garlic; but that is Nature's fault, not the author's. There they stand, Lilium and Allium, sections of the extensive and beautiful Lily family, and included with them are plants of the most diverse character Squills and Saffron, Asphodel and Tulip, Snake's-head and Star of Bethlehem, Lily of the Valley and Solomon's Seal, even such oddities as Asparagus and Butcher's Broom—plants that have given up producing leaves. In some of these it is not easy at first sight to see any resemblance to the typical Lily; yet a closer examination of the flowers will discover the fundamental plan of the Lily even in so remarkable a plant as Butcher's Broom.

Now, what are the family characters whereby such differing forms may be reconciled? Taking the flower first as the most important part, it is a six-parted perianth, like those we found in the Amaryllis family, all the segments resembling petals, but in this case the ovary is within the flower (*superior*), not below it as in those. There are six stamens, and one or three styles. It is rarely that the stem takes on a shrubby character, but it does approach to this in the case of Butcher's Broom alone among our native species. The rootstock is either a bulb, or it is thickened, fleshy, and creeping.

So much for the general characters; now let us consider the only species of the typical genus, which by the way is merely a naturalised plant found in no other British locality but Mickleham, in Surrey. This is the Purple Martagon Lily (*Lilium martagon*), with flowers two or three inches across, hanging in an inverted position, with the petals curled up round the flower-stalk and away from the stamens and style. The red-brown anthers are attached to the filaments by their centres, and are so loosely hinged that the slightest movement of the air swings them this way or that. Such an anther is described technically as *versatile*; if an insect touches either end, it swings round and lays its whole length against the insect. Only long-tongued insects can get any good out of this flower in the shape of honey, and Müller, who has had better opportunities of observing this plant than falls to the British botanist, declares it is adapted for cross-fertilisation by Hawk-moths,

Petal and Nectary of Martagon Lily

who are assisted by *Noctuæ*. On this account its sweet perfume is only given off at night. The flower being so widely opened, honey would in the ordinary way be accessible to all callers, but to save it for the long-tongued ones who can cling to stamens and pistil and dust themselves with pollen what time they are drinking, an ingenious plan is adopted. The honey lies along the centre of the basal half of each petal, in a deep groove which is walled in by thick, arching borders fringed with stiff hairs, which preclude admission to any insect's tongue unless it be inserted at the open end. The accompanying figure (after Müller) will make this plain. After the moth has clung to the stamens in alighting, it climbs up to the tips of the petals, and feeling the way between the little pointed warts studding the surface of the petal, finds the entrance to the nectary (*e*).

If we were to dig up the bulb of this Martagon Lily—but I do not suggest that any reader who comes across its sole British station should do so, for a White Lily or Tiger Lily from the garden will do as well—we should get the best insight into the true character of these bulbs. In all species of the genus *Lilium* the bulbs consist of a large number of fleshy scales overlapping each other. Now, the very form of these will convince the investigator that they are merely undeveloped leaves, which have been enormously thickened to serve as storehouses for the starchy wealth of the plant. Any one of those scales separated from the mass will develop into one or more tiny bulbs. Sometimes such miniature bulbs form at the base of the scale, sometimes around the edge.

I have had bulbs of the beautiful Japanese Lily (*L. auratum*) into which slugs had bored by eating away the stem, and so hollowing out the bulb; to all appearance the bulb was ruined, but by cleaning out the slugs and replanting, I got some forty or fifty tiny bulbs produced by the sound scales. A big bulb will sometimes deliberately break itself into several smaller ones, and there is always a tendency to go away from a centre. Thus, if a Lily-bulb be planted say in the centre of a tub or large flower-pot, it will be found after two or three years that the plant no longer comes up in the centre, but that several smaller stems shoot up from near the circumference. This phenomenon is due to the exhaustion of the soil for that species in the centre, and the plant makes an effort to get away from it. If we are not merely plant-growers, but investigators also, we shall want to know how it is done, and on carefully taking off the top-soil we shall be enlightened. The old bulb has broken up into five or six smaller ones, and these instead of sending up vertical stems, have shot out horizontally, until the pot or tub has stopped their progress, when they grow upwards. The scales of these bulbs become thin and flabby by their material being used up for the new stems and flowers, but when the material freshly elaborated by the leaves is carried down in autumn, it goes not to the former station for the bulb, but only to the point where the stem took its upward direction. Thus is the mysterious movement of Lily-bulbs and Tulip-bulbs explained. This and several other facts we have already adduced in these pages suggest that if Alexander Pope returned to this life he might desire to modify his couplet in the *Essay on Man*—

> "Fixed like a plant on his peculiar spot,
> To draw nutrition, propagate, and rot."

I have started with one of the most highly developed members of the family, but it is clear from the evidence still left to us in our flora that the family had quite a humble origin in plants with poor little green honeyless flowers, from which condition the attention of various insects have induced the plants, in order "to merit a continuance of such esteemed patronage," as shopkeepers say, to improve their flowers on certain differing lines, which has led to the variety we now find—the family having in its worldwide distribution no less than 187 branches or genera with something like 2500 species. If we examine the inconspicuous and little-known flowers of the Black Bryony (*Tamus communis*)—which is not included in the Lily family—it may give us an idea of the kind of flower possessed by the founder of the family. This flower is either male or female, never including both stamens and pistil. The green perianth consists of six segments in two series, assuming a bell shape, and upon each of these segments is fixed a stamen with the anther turned toward the centre of the flower. Below the female flower there is a three-celled ovary, with three short styles and two-lobed stigmas, and this develops into the brilliant red oblong berry which hangs in clusters from the hedges in autumn when the handsome heart-shaped leaves have become bronzed almost to blackness. Plants of this family are among the oldest known as fossils, and we know from the universal distribution of the Lilies throughout the world that they too are of great age.

Now turning to those plants admitted within the Lily family, we have two species constituting the Asparagus tribe in which the conditions of the flowers are very similar to those of Black Bryony, except that the ovary is here *within* the female flower instead of below it. Asparagus (*Asparagus officinalis*), whose medicinal virtues were well known to the Romans at least two hundred years B.C., and are still recog-

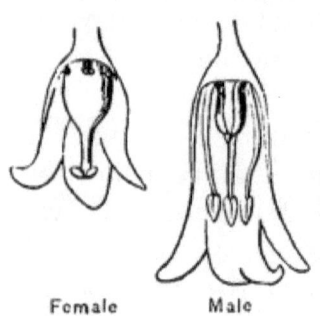

Female   Male
Asparagus

nised, is found in a few localities on the coasts of Dorset, Cornwall, and Wales, as well as in one Irish locality and the Channel Islands. It is, however, best known as a cultivated plant. The leaves are reduced to minute triangular scales scarcely noticeable, and in their axils are branches of needle-shaped branches, or "cladodes," which generally pass for the leaves. The tiny flowers are bell-shaped, whitish or yellowish veined with red, and produced either singly or in pairs. As in Black Bryony, the stamens are borne by different flowers from those containing pistils, but it would appear to be almost certain that this species is descended from one in which the sexes were combined in each flower, for the male flowers contain a rudimentary pistil, and the female flowers have rudimentary stamens. The male flowers are much larger than the females, an evident purpose of this being to induce a visiting insect to enter the male flowers first and get dusted with pollen to be afterwards distributed among the female flowers. In spite of their lack of colour, the flowers cannot be said to be inconspicuous, and they are more easily found by a pleasant odour which evidently gives the bees confidence in the matter of honey, which they duly find on dipping their tongues into the lower part of the flower. The Hive bee, the Leaf-cutting bee, and several other of the smaller species are the chief patrons of Asparagus, some of them being merely pollen - collectors, but having collected in the male flowers, they visit the females under the impression that they can add to their load there, but they leave a trifle instead, and effect fertilisation. There is no bulb to Asparagus, but the creeping rootstock is thick and fleshy. Similar creeping rootstocks are found in Butcher's Broom, Solomon's Seal, Lily of the Valley, Bog Asphodel, and Herb Paris; so that it must not be taken for granted that all the Lily family have bulbs. It is not difficult to understand the evolution of both corms and bulbs from such rootstocks.

The Butcher's Broom (*Ruscus aculeatus*) is another rare plant found naturally only in the southern half of England, but usually in some quantity where it does occur. It is an evergreen shrub, as already stated, and like Asparagus it has given up developing true leaves— why, is not clear. We may surmise that at one period in its history its leaves and stems were

Butcher's Broom

so persistently browsed down that the former were of no use to it, so it took to flattening out short branches until they resembled leaves and exercised their functions, but had the advantage of being leathery and indigestible. Colour is given to this supposition by the needle-point at the tip of the oval cladodes, evidently designed to prick the muzzle of any herbivorous creature attempting to do violence to the plant. A young shoot shows the true scale-like leaf at the base of the cladode, but this soon shrivels. Different shoots—sometimes separate plants—produce the male and the female flowers, and these may be at once detected by the fact that the cladodes bearing female flowers are broader than the others. As in Asparagus, these are succeeded by brilliant red berries, which are very conspicuous against the dark background afforded by the cladode.

The Lily of the Valley (*Convallaria majalis*) is still found in a wild state in some of our woods. All its leaves come direct from the creeping rootstock, two or three of them with their bases sheathed one in the other. The broad bell-shaped nodding white flowers are borne on an arching scape. The organs are all contained within the perianth, the stamens close round the style with the anthers able to drop their pollen upon the edges of the stigma if cross-fertilisation has not taken place soon after the flower opens. I fear this plant must be included in the list of vegetable frauds, for by means of a delicious odour it creates the hope and belief among certain bees that it provides nectar for friendly callers. This appears to be an error, yet the hive-bee constantly goes on that errand, and has to content herself with pollen,

Lily of the Valley

which is always handy in the hive, and in the collecting of it she must frequently effect cross-fertilisation. Red berries succeed the fertilised flowers.

In Solomon's Seal (*Polygonatum multiflorum*) there is a distinct leafy stem, two or three feet in height, arching over with the leaves arranged in two rows, and the flowers hanging by slender foot-stalks from the axils of the leaves. The pollen of Lily of the Valley is protected by the drooping of the bell-shaped flower; here the greenish-white tubular bells have similar protection afforded by a like habit with the additional security furnished by the leaves above them. The mouth of the flower is dilated to enable humble-bees to put their heads inside whilst they are reaching for the honey on the floor (or is it roof?) of the flower. The stamens are attached to the tube about half-way up, and the bee has to push his tongue between the stigma and the anthers in order to reach the honey. Whilst one side of his head is coming in contact with the stigma, the other side is taking up a new supply of pollen. It is, of course, a matter of chance whether the right or the left side of the style will be selected by the bee, but in a number of visits both sides are sure to be covered with pollen early, and afterwards each visit would result in cross-fertilisation.

The Onion tribe is fairly well represented in our flora, the only common species being the Ramsons, or Broad-leaved Garlic (*Allium ursinum*), whose leaves are often mistaken for those of the Lily of the Valley until they happen to get trodden upon or otherwise bruised, when the veriest tyro in matters botanical knows they have close affinity with the pungent

Onion, beloved of cooks. The segments of the flower are white, and in this case they spread widely from the base, so that each blossom is a six-rayed star, and about a dozen of them are united in an umbel at the summit of a three-sided scape. Before the flowers open, the entire umbel is wrapped up in a two-leaved spathe. The ovary secretes honey which is retained between the carpel-divisions and the base of the three inner stamens, and to obtain it the insect must thrust its head between stigma and anther, touching both, and so effecting cross-fertilisation much as in Solomon's Seal, though the shape of the perianth is so different. When the flower first opens, neither the stamens nor the pistil is mature. First the anthers mature and open one after the other, the inner set first. They open on the side towards the style, then turn their pollen-covered faces upwards to offer a greater obstruction to any insect alighting. It is not until they have all burst that the stigma is mature, and should it not quickly be dusted with pollen brought from another plant it curls over until it touches one of the latest anthers and so secures fertilisation. A Humble-bee (*Bombus pratorum*) has been observed to visit the flowers.

The Wild Leek (*Allium ampeloprasum*) is regarded as only naturalised in a few localities, but I have found it growing on Cornish cliffs and banks in such situations as preclude the probability of its having been planted or escaped from gardens far away. This plant has rounded leaves two or three feet in length, and the flower-scape rises to a height of four or six feet with a globular head of buds wrapped in the single spathe which ends in a long upright beak.

In one variety of this species only a few of the flower-buds open, the majority of them developing into bulbils. Various insects assist in the work of fertilising both species.

The pretty little Grape Hyacinth (*Muscari racemosum*), found wild only in the eastern counties, but frequently naturalised near gardens, has its segments united throughout their length into a globose cylinder with six little lobes round its mouth. The stamens are attached to the walls of the perianth about half-way up. Bees visiting the flower bring their heads in contact with the simple stigma and then push their tongues against the anthers, shaking down the pollen. Bees are attracted by the plum-like odour, which has also been regarded as musk-like—hence the name *Muscari*.

The Vernal Squill, or Sea Onion (*Scilla verna*), which gems the short turf of headlands on the West coasts with its bright-blue starry flowers, has a similar bulb to that of the well-known garden Hyacinth, and long narrow concave leaves. The fragrant flowers are clustered on one or two short scapes, and have their segments separated to the base. A similar flower, but of reddish-purple hue, is produced by the Autumnal Squill (*S. autumnalis*), on long scapes. A noteworthy difference is evident between these species. *S. verna* produces its leaves before the flowers in spring, and *S. autumnalis* flowering in summer does not produce its thin half-round leaves until autumn. This is no doubt due to the different flowering period. *S. verna*, drawing upon its savings of last year, sends up its leaves early, and as soon as these are out the flower-scape follows;

the leaves working for bulb-renewal at the same time as the flowers are expanded. *S. autumnalis* flowers in July and August, when the pastures and rocky places it affects are dried up. This is not a good time for putting out its thread-like leaves, so it waits until the flowering is all over and the autumnal mists offer a better opportunity for their work. One other species of Squill is the favourite Blue-bell, or Wild Hyacinath (*S. nutans*), which differs from the others by its larger size and elongated bell-shape, the segments being united near the base. Both leaves and flowers are produced in spring. It is interesting to note the behaviour of the flowers in their various stages. When the scape appears from amid the leaf-rosette, the buds all stand erectly, and so continue until they open in turn, the lowest first, when they fall over on their foot-stalks and hang down to be out of the way of the flowers that yet need fertilisation. When fertilisation is completed, and the seeds are forming, the fruit begins to assume the position of the unopened bud, because the ripe capsule opens at the top. It would be too late for the capsule to erect itself after the seeds were ripe or nearly so, for by that time all the tissues have become dry and hard. The scape has lengthened considerably after flowering, and each of the capsules has become inflated and parchmenty. The small black seeds are loose within it, and as the wind blows or some animal brushes past, the scape springs back and forth with a jerk that sends a number of the seeds to a considerable distance.

Of the three species of Star of Bethlehem, two have become naturalised in our copses here and there, but only the Spiked Star of Bethlehem (*Ornithogalum pyrenaicum*) is native, though it is of very restricted range over a few counties. In the neighbourhood of Bath it grows abundantly, and the young shoots are sold in Bath market in spring as French Asparagus, and eaten instead of the true Asparagus. The bulbs of *O. umbellatum*, the Star of Bethlehem, are cooked and eaten in Palestine. These plants still retain the ancestral green on the outside of the perianth, and a streak of the same colour on each segment inside, looking as though on the way to becoming white as in Ramsons. The Yellow Star of Bethlehem (*Gagea lutea*) belongs to another genus; it is fertilised by bees, but open only in the morning. The stigmas ripen before the anthers.

The neat and pretty Fritillary, or Snake's-head (*Fritillaria meleagris*), has adapted itself to the visits of bees by overlapping the edges of its perianth-segments without soldering them together, and by colouring them in a chequered pattern of pink and purple. Into this drooping bell the bee flies, alighting on the stigmas, which are brushed by its abdomen as it climbs over the anthers to reach the honey, secreted by glands at the base of the ovary. This passage of the anthers loads the hairs of the abdomen with fresh pollen for the fertilisation of other flowers.

Our rare solitary species of Tulip (*Tulipa sylvestris*) is coloured bright yellow. It is larger than the Fritillary, but of similar form, except that the segments of the flower have their tips curved back. They attract bees by their fragrance, but secrete no honey, fertilisation being effected by the bees as they collect or eat pollen.

The much rarer Alpine Tulip (*Lloydia serotina*) has retained the more open, star-shaped blossom of white or yellowish hue, with purple veins to show the way to the nectaries. The honey is secreted by a thick ridge at the base of the perianth-segments and along their middle, easily accessible to the short-lipped insects, chiefly flies, by which it is fertilised. The inner series (three) of stamens ripen and discharge their pollen before the stigmas are mature, which favours cross-fertilisation. In some specimens the stigma is on a level with the anthers, and here self-fertilisation is favoured. The plant is especially interesting as a probable remnant of the flora of the Great Ice Age, its stations in this country being a few rocky ledges of the Snowdon range, and elsewhere it is found abundantly close up to the line of perpetual snow.

The Meadow Saffron (*Colchicum autumnale*) looks uncommonly like a Crocus, but whereas the Crocus family are all content with three stamens and one style, the Saffron has six stamens and three styles, which testify to its Liliaceous character. It has a corm, too, not like that of the Crocus, but of different shape, and covered with shining scales of a chestnut colour. Its flat, narrow, lance-shaped leaves are not Crocus-like, but its pale-purple flowers are, the ovary lying within the corm, and the thread-like styles reaching up therefrom. The yellow stamens are attached to the sides of the expanded portion of the flower, and the honey is secreted by their basal portion, and runs into grooves in the perianth, where it is protected by fringing hairs. The stigmas mature before the anthers, so that the first chance is given to cross-fertilisation, when the Humble-bees, such as *Bombus hortorum*, visit it. Should their visits not take place until after the anthers have burst, there is a possibility of self-fertilisation taking place; but every effort is made to avert this, the anthers turning on their hinges so as to place the pollen-covered surface as far away from the stigmas as possible. In order to reach the honey the bees have to creep right into the flower, and in so doing to bring their head and fore-legs into contact with stigmas and anthers. Should the latter be ripe, they take away a liberal load of pollen for the next flower.

The Bog Asphodel (*Narthecium ossifragum*) before it flowers looks like a miniature Flag, its leaves being of the same sword shape. The flowers are much on the Garlic plan, but coloured bright golden yellow within and greenish without. The Scottish Asphodel (*Tofieldia palustris*) is very similar, but the flowers are much smaller, and yellowish-white in colour; these are not very conspicuous, but the provision of honey by the carpels attracts short-tongued insects, and in default of their attendance there is a good chance of self-fertilisation owing to the anthers and stigmas maturing together. Both these plants have a creeping rootstock, instead of a bulb.

The sole remaining tribe of Lilies to be glanced at consists of Herb Paris (*Paris quadrifolia*), a local plant of the woodlands. It is placed at the end of the family because it is an anomaly. Its creeping rootstock sends up a simple stem with a whorl of four large broadly-oval leaves, above which is the solitary flower, consisting of from three to five (usually four) long lance-shaped sepals, as many narrower yellow petals, from eight to twelve stamens with the connective extended in a slender point far beyond the anthers, and a large, almost globose dark-purple ovary with four styles of the same colour. The ovary and styles though perfectly dry shine as though wet, and the flower exhales a disagreeable odour. There is no honey, but the vile odour and the moist-carrion appearance of the ovary attracts the Dung-fly (*Scatophaga merdaria*) and other two-winged flies, which alight on the ovary and lick it, so well are they deceived by appearances. Failing to get satisfaction there, they turn to the stamens and climb up the stiff filaments, dusting their under-sides with pollen as they pass the anthers. The object of the extended connective is to enable the flies to crawl right over the anthers, so that on visiting other flowers they may be well primed with pollen. To assist in cross-fertilisation, the stigmas are already mature when the flower opens, but the anthers do not shed their pollen until several days later.

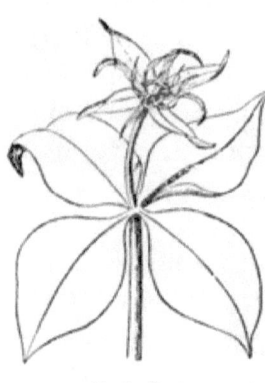

Herb Paris

RUSHES AND REEDS

MOST of the plants to which we must now devote a few pages may be best described as degenerate Lilies—plants that have turned their backs upon the insect friends that have induced so many of the family to become brilliant, gaily-coloured, and often sweet-scented. They appear to have argued that expensive petals richly coloured, with their adjuncts sweet perfumes and sweet liqueurs, made too heavy a demand upon their resources; they must retrench. They have retrenched to such an extent that all-round degeneration has resulted—simple stems, leaves scarcely recognisable as such, and poor little brown or green flowers massed together in cymes, and fertilised by the winds. They have been edged out of the richer lands, and have had to crowd together in the poor soil of bogs and marshes, sandy seashores, or moist places on the moorland. Yet, in spite of their degeneration, they have clung to the evidences of their distinguished ancestry: you have only to examine their flowers—and these have become so small you must do it with a lens—to be satisfied of their relationship to the magnificent *Lilium auratum*, the gaudy Tulip, and the pungent Onion.

The Wood-rushes (*Luzula*) have kept nearest to the Lily type in their long flat leaves clothed with long silky hairs. Their flowers vary from one-fifth to one-twelfth of an inch across, and are of harsh texture; but there are the six segments, the six stamens, and the three stigmas. It is true the three cells of the ovary have been reduced to one, but the capsule still contains the three seeds, and opens by three valves. In adapting themselves to wind-fertilisation, these plants have had to make their stigmas long and thread-like and covered with little raised points (*papillæ*) all over to give them a better chance to catch the dry pollen-grains that fly from another plant. Looking at this figure of a Wood-rush flower, the reader might object that the stigmas by bending down could come into contact with the anthers and so effect self-fertilisation without calling in the aid of the wind to effect a cross; but self-fertilisation is prevented in a very simple manner—the stigmas are ripe, and as a rule have been fertilised, before the anthers open.

Wood-rush

In the Rushes (*Juncus*) the flower-parts are, popularly speaking, much the same as in the Wood-rushes, but the ovary is usually three-celled, and the seeds are many instead of three. The three outer segments of the perianth are keeled, and all are more slender. The leaves and stems are smooth, and the leaves in some cases reduced to mere sheaths round the stem. The pith from the stems of these plants constituted the wicks of the old "rushlights." In Bulrush (*Scirpus lacustris*) the segments of the perianth have been reduced to six bristles, and in the related group, the Sedges (*Carex*) the perianth has disappeared completely, the female flowers being enclosed instead in a bottle-shaped sack called the *perigynium*, which is composed of two united chaffy scales known as *glumes*. These glumes are of similar nature to the bracts and spathes we have met in some preceding families, and we shall meet them again, or similar bodies, in the Grasses.

The mention of spathes reminds me of a notable example of such, which does not properly come within the subject-matter of this chapter —nor, for the matter of that, of any other chapter, so it may as well be mentioned here as elsewhere. I allude to the well-known Cuckoo-pint, Wake Robin, or Lords and Ladies (*Arum maculatum*). Everybody is well acquainted with the remarkable cowl-like "flower" of this plant, which is really a large assemblage of flowers within a spathe, all of which have so completely got rid of their perianths that they have had to cover their nakedness in this huge tent of a spathe. Few of the many that make efforts in early spring to grub up this strange plant from hedge-bank and copse, pause to consider the interesting series of

Cuckoo-pint

episodes by which fertilisation is effected, and how the plant stoops to deception of a systematic character in order to compass its ends. The handsome, large, arrow-shaped, and spotted leaves are in evidence for at least a month before the flowers. They are of such soft texture that they would soon be eaten by herbivorous mammals, but that the plant has learned by experience in the past, and now develops such acridity that the first bite is enough for any creature. I believe that the dark-purple spots on the leaves are a warning sign, but it is singular that in some districts these spots are never developed. Within the centre of the cowl stands a thick, fleshy, club-shaped organ, more or less of a dull purple colour, and if we tear off the front lower portion of the spathe we shall find the lower half of this club (*spadix*) to be crowded with minute flowers. Examination with a simple lens resolves these into three distinct sets of organs ranged one above the others round the spadix.

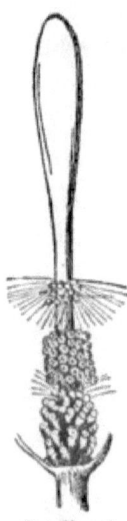

Spadix of Cuckoo-pint

The lowest series consists of stigmas, the next of anthers, and the uppermost series of stamens that have become converted into long oblique hairs which radiate to the walls of the spathe. Now, under the old theory that flowers exist solely to give pleasure to the human eye and nose, this arrangement and specialisation is absurd and uncalled-for; but in the light of the doctrine that teaches the utility to the species of the shapes, colours, odours, and special arrangements of the flowers, everything about the Cuckoo-pint becomes easy of comprehension, transparently clear in fact.

Whenever a flower exhibits the liver-like hue we see in the spadix, accompanied by an unpleasant odour, we may safely conclude that it is seeking to attract carrion-loving flies to perform a service for it under false pretences. Before it was clearly seen how important a part insects play in the fertilisation of flowers the relative position of anthers and stigmas in Cuckoo-pint would have been considered an admirable arrangement to secure fertilisation; but when all the circumstances are known this is seen to be impossible, and were it not for the visits of numbers of a small fly (*Psychoda*), the Cuckoo-pint would cease to produce berries.

The stigmas ripen and pass maturity before the anthers shed their pollen, therefore if fertilisation is to be effected at all it must be by pollen brought from an older flower. The flies that act as carriers of this pollen are very small and moth-like, feeders upon decaying fungi and similar delicacies. To describe their *modus operandi*, we will suppose a few dozens or hundreds of them have already been in the spathe of a Cuckoo-pint that flowered a few days earlier than the specimen we are considering. The stigmas mature, and the "flower" evolves a most unpleasant ammoniacal odour, accompanied by a rise in temperature. The *Psychoda* fly to the purple spadix, and finding out they have been fooled as to its real nature, they follow the scent which leads them to the lower regions. The barrier of hairs across the entrance easily admits insects going downwards—on the *facilis descensus averni* principle—though it prevents return. This barrier passed, the flies are imprisoned for the present, and can do nothing but creep about and vainly fly up towards the light, as we can see by gazing through the screen of hairs. But in these attempts at flight they are shaking off the pollen they brought from the other flowers, and some of this falls upon the visited stigmas and fertilises the seed-eggs. When these organs are no longer susceptible, each excretes from its centre a clear drop of nectar, that the flies may drink, and, forgetting their enforced imprisonment, go away with pleasant recollections of the Cuckoo-pint.

By the time the refreshments are exhausted the anthers have come to maturity, and shed their pollen in showers upon the flies, who finally get well covered with it. Then the hairs of the barrier shrivel up and leave a wide exit up which the *Psychodæ* can fly, and with still fresh memories of the nectar, they seek the freshest Cuckoo-pint in the vicinity to repeat the experience, like topers going from tavern to tavern. Now, it will be seen that this provision is a piece of pure altruism in the interest of the *species*, not the individual. The individual's turn was served when the flies brought the pollen from an older Cuckoo-pint and fertilised the stigmas. The loading of the flies with pollen, and sending them away with the *pleasures* of their incarceration uppermost in recollection, is dictated solely in the interests of the species as a whole.

Where the rare Sweet Flag (*Acorus calamus*) grows, it is interesting to compare it with the Cuckoo-pint. Its leaves and stem are very Flag-like, and as it grows on the margins of ponds and streams, it may easily be overlooked as such. The flowers, to a very large number, are crowded round a spadix, but the spathe instead of enveloping them is long and leaf-like, and looks as through it were a mere continuation of the scape. Here each flower has its own perianth, which is six-parted, and includes within its shelter six stamens and a two- or three- celled ovary.

GRASSES

WHAT is to be said of the Grasses, the commonest plants of all that grow, that are daily trodden under foot, and that everybody knows are uninteresting except as the material of which tennis-courts, cricket-pitches, and golf-links are made? The difficulty is really to construct a few brief paragraphs that shall give some general idea of the family so far as it is represented by the hundred and twenty species that occur in these islands. Over three thousand species are known from all parts of the world, and so greatly do these differ among themselves, though agreeing in all the principal points of structure, that the family has had to be divided into about three hundred genera, and these in turn grouped in tribes, and the tribes marshalled into series. Our own six-score of species represent no less than forty-eight genera. Although the Grasses are common to all climates and all parts of the globe, it is only in Temperate regions that they form continuous lawns and pastures. We cannot boast of giant Grasses in the shape of Bamboos sixty or seventy feet high, and nothing larger than the Great Reed (*Phragmites communis*), yet we have in our smaller native species a beautiful and perennial green mantling of hill and valley, whose nutritious herbage is indirectly of the highest value to a flesh-eating race, apart from its æsthetic importance.

The roots of Grasses are always fibrous, and they take hold of the soil so completely that they bind it into a matted turf; some of the species have also creeping stems which run just below or just above the surface, rooting as they go, so that in a well-established pasture there is no part of the surface soil that is not in almost immediate contact with root or stem of grass.

Where the Marraw, or Sea Reed (*Ammophila arundinacea*), grows upon our shores, it often does good service, its great, fleshy, creeping stems holding the sand together, and not only building up by this means a barrier against the force of the waves, but also preventing the sand from being blown farther inland to do damage. It is said that the town of Hull is indebted to this grass for its continued existence, otherwise threatened by the encroachments of the sea—Marraw having given strength and solidity to the sandbank known as Spurn Point. In the Hebrides and on the French coast also the plant has been actually planted on sandy shores for the purpose of reclaiming level tracts from the sea. So well has this valuable character of the Marraw been appreciated, that in our own country its destruction has, in former times, been prohibited by Act of Parliament.

The characteristic structure of all Grass stems is that of a series of straight tubes, placed end to end, the joints being more or less swollen and solid. Familiar examples of this structure on a larger scale may be seen by examining a Bamboo rod or a length of Wheat or Barley straw—all Grasses. It is a singular fact that in growing Grasses we never see the stem with the exception of the top joint, which bears the flowers. This is due to the peculiar formation of the leaf-stalk, which is flattened out until it is broader than the leaf-blade, and then wrapped around about two joints of the stem.

Grass spikelet

Taking a growing stem of Grass and pulling off a leaf carefully, you will find that it is not joined where it appears to be, but at some distance lower down. These leaves are what a botanist would term *linear*—that is, with more or less parallel sides, and very narrow in proportion to their length. Their veins always run parallel with each other and with the midrib.

A variety of the Common Oat-grass (*Arrhenatherum avenaceum*) has become a bulbous plant in order to avoid being dried up in summer on the poor clayey or sandy soils it affects. At the base of its stem it produces a cluster of little onion-like tubers, stored with food and capable of rapid growth when rain again falls. This character has earned for it the name of Onion Couch, the last word having reference to the fact that it is as troublesome as

Couch-grass (*Agropyrum repens*) when it gets hold of garden or cornfield.

The flowers are variously grouped in spikes, racemes, or panicles, and it is not easy for a tyro used only to the larger flowers of Roses, Orchids, Lilies, etc., to make out their structure. The surest way to understand a Grass-flower is to analyse it, taking a spikelet and pulling it to pieces. A spikelet is what appears to be a single flower, and it must be understood that its parts vary in the different genera—both in the number of its envelopes and of the real flowers it contains. The envelopes are chaffy scales corresponding to the bracts and spathes of ordinary plants, but here they are called *glumes*—the Latin word for chaff or husks.

Usually a spikelet contains a pair of flowerless glumes enclosing the boat-shaped flowering glume and a flat scale (*palea*). Within the flowering glume are two little colourless scales called *lodicules*, representing the perianth of the Lily, and the essential organs; these are a one-celled ovary, with two long plumy stigmas and three stamens. As all Grasses are fertilised by the wind, the stigmas stretch far apart in order that they may the better catch the flying pollen-grain on their sticky feathers. The stamens consist of a long hair-like filament, upon which a two-celled anther is so beautifully poised by its middle that as it hangs out of the flower it sways in the slightest breath of air and shakes out its pollen to the breeze.

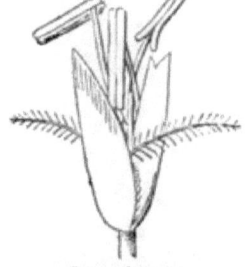

Grass-flower

Owing to the fact that Grasses grow together in such enormous numbers, cross-fertilisation by such means is certain; but in addition a little pollen-carrying is done by certain small flies. The seed is a hard grain, of which Wheat, Barley, and Oat are familiar examples. It is frequently wrapped closely within the palea, requiring the operations of the mill to separate it in those grains used for food.

INDEX

Agrimony, 45; its hooked calyx, 45.
Agrimony, Hemp, the simplest native Composite, 177; its attractiveness for butterflies and bees, 179.
Alder, 295.
Anemone, Wood, 68; fragrant but honeyless, 69.
Apple, 33; honeyed flowers, cross-fertilised, 33; structure of fruit, 34; dispersal of seeds, 35.
Asparagus, two forms of flowers, 328.
Asphodel, Bog, 337; Scottish, 337.
Avens, Wood, 44; its hooked nutlets, 45.
Azalea, Trailing, 217.

Balm, 273.
Balsam, 134.
Balsam, Yellow, 135.
Barren Strawberry, 39.
Bartsia, Alpine, 258; Red, 258; Yellow, 258.
Beam, White, 32.
Bearberry, 213.
Bedstraw, Field, 174; Hedge, 173; Lady's, 173.
Beech, 295.
Bell-flower, Ivy-leaved, 206; Nettle-leaved, 205; Spreading, 205.
Bennet, Herb, 44.
Betony, 280.
Bilberry, 209; remarkable stamens, 209; their action, 210.
Birch, 295.
Bird's-foot Trefoil, 146.
Blackberry, 46, 48; its climbing hooks, 48; great variation, 49.
Black Bryony, 327; prototype of Lily flowers, 327.
Black Horehound, 283.
Blackthorn, 50.
Bladderwort, 267, 270; trap of, 270.
Blue-bell, 334.
Bluebottle, 190.
Blue-eyed Grass, 312.
Bog Asphodel, 337.
Bogbean, 229.
Bog Whortleberry, 211.
Borage, 237.
Box, 287; obnoxious to insects, 290; male and female flowers, 290.
Bramble, 48; Stone, 46.
Brooklime, 250.
Broom, 150; mechanism of flower, 151.
Broomrapes, 262; leafless parasites, 263.

Bryony, Black, 327.
Bugle, 273, 285.
Bugloss, Common, 239; Viper's, 235, 236.
Bulbous Buttercup, 65.
Bullace, 50.
Bulrush, 341; its petals reduced to bristles, 341.
Burdock, 189; its hooked fruits, 189.
Burnet, Great, 56; Salad, 55.
Butcher's Broom, 323; its branches made to serve as leaves, 329.
Butterbur, 186.
Buttercups, 62, 63, 64; Bulbous, 65; Creeping, 65.
Butterwort, insectivorous, 267, 268, 269.

CABBAGE FAMILY: anti-scorbutic properties, 89; pungency and acridity protective, 89; Wild Cabbage, 91; Red, White, and Cow Cabbages, 91; many garden vegetables from one ancestral form, 91.
Campion, Bladder, 106, 108; Red Campion, 104, 105, 106; Sea Campion, 107, 108; White Campion, 105; Campion Moths, 107.
Candytuft, 93; its odd-sized petals designed for more effective advertisement, 93.
Carline Thistle, 189; everlasting and hygrometric, 190.
Carnation, 108.
Carrot, Wild, 164, 169.
Catchflies, 116; sticky and absorbent hairs, 116; waste turned to profit, 117.

Cat-whin, 149.
Celandine, Greater, a Poppywort, 80.
Celandine, Lesser, a Buttercup, 57, 58; fleshy roots or treasure-bags, 59; Culpepper on, 59.
Celery, Wild, 169.
Centaury, 233.
Chamomile, 182; Wild Chamomile, 183.
Cherry, Bird, 51; Dwarf Cherry, 51, 52; Gean Cherry, 52.
Chickweed, 113, 114, 115; a species that has seen better days, 114; Mouse-ear Chickweed, 111.
Chicory, 196.
Cinquefoil, 37; Marsh Cinquefoil, 39.
Clary, 278.
Cleavers, 174; its flinty hooks, 174.
Clematis, 72; its feathered stigmas and fruits, 71, 72.
Cloudberry, 47.
Clover, 152; Dutch Clover, 153; Clover, humble-bees and mice, 153; Red Clover, 153; fertilised exclusively by humble-bees, 156; White Clover, 153, 154.
Coltsfoot, 186.
Columbine, 74.
Corn Cockle, 110.
Cornflower, 192.
Cornish Moneywort, 253.
Corn Marigold, 184.
Couch, 348.
Cowberry, 211.
Cow Parsnip, 165; associated flowers, 165; oil tubes, 166.
Cowslip, 221.

Cow-wheats, Yellow and Purple, 254.
Crab, Wild, 35.
Cranberry, 211.
Crane's-bills, 124; Bloody Crane's-bill, 126; Meadow Crane's-bill, 124; Mountain Crane's-bill, 126; Shining Crane's-bill, 128; Wood Crane's-bill, 124.
Creeping Jenny, 223.
Crocus, Autumnal, 312; Vernal Crocus, 312.
Crossworts, 85, 173.
Crowfoots, 64; Water Crowfoot, 65.
Cuckoo-pint, 341, 342, 343, 344.
Cypress Spurge, 286.

DAFFODIL, 319.
Daisy, 179; its two hundred florets, 180; ray-florets and disk-florets, 182; Ox-eye Daisy, 184.
Dandelion, 196; its strap-shaped florets and their terminal teeth, 197; its seed and parachute, 198.
Dead-nettles, 273.
Dropwort, 54.
Dyer's Greenweed, 148.

ELM, Common, 291; Wych Elm, 291.
Euphrasy, 259.
Eyebright, a root-parasite, 259

FEATHERFOIL, 228.
Felwort, 230.
Feverfew, 184.
Field Madder, 171.
Figworts, wasp-flowers, 253.

Flag, Yellow, 311; its puzzling flowers, 315; Sweet Flag, 344.
Flea-banes, 182.
Forget-me-not, 239, 240; its hooked fruits, 240; Field, Wood and Water Forget-me-nots, 240.
Foxglove, 242; how the bee works a flower-spike, 243; three stages of the flower, 244.
Fritillary, 335.
Furze, 149; its explosive flowers and pods, 149.

GARLIC, Broad-leaved, 331.
Gentian, Alpine, 231; Field Gentian, 229; Marsh Gentian, 230; Spring Gentian, 231; Yellow Gentian, 231.
Geraniums, 123; Dove's-foot, 127; Round-leaved Geranium, 127; Small-flowered Geranium, 127.
Gipsywort, 275.
Gladdon, 317.
Globe-flower, 66.
Goat's-beard, 199.
Goldielocks, 63.
Goosegrass, 171, 174; how it climbs, and distributes its seeds, 174.
Grape Hyacinth, 333.
Grass, Blue-eyed, 312.
Grasses, 346.
Great Reed, 347.
Ground Ivy, 279.
Ground Pine, 286.
Groundsel, Common, 188; Mountain Groundsel, 188; Stinking Groundsel, 189.
Guelder Rose, 176.

HARDHEADS, 191.
Harebell, 202; the reason for its inverted flowers, 204.
Hare's-ear, 167.
Hawkweed, 196; Mouse-ear Hawkweed, 199.
Hawthorn, 37.
Hazel, 295.
Heath, Cornish, 214; Cross-leaved Heath, 213; Fringed Heath, 214; Irish Heath, 215; Purple Heath, 213; St. Dabeoc's Heath, 216.
Heather, 216.
Hellebores, 67; petals converted into honey-flasks, 67.
Hemlock, Water, 169.
Hemp Agrimony, the simplest Composite, 177, 178.
Hemp Nettle, 280.
Herb Bennet, 44; its hooked fruits, 45.
Herb Paris, 329, 337, 338.
Herb Robert, 127.
Hollyhock, 118.
Honeysuckle, 177; its flowers adapted for long-tongued insects, 177.
Hop, 291, 294.
Hornbeam, 295.
Hyacinth, Grape, 333; Wild Hyacinth, 334.

IRIS, Yellow, 315; Fœtid Iris, 317.
Ivy-leaved Toadflax, 263.

JACK-BY-THE HEDGE, 90.
Jewel-weed, 135; its irritable capsules, 135.
Juniper, 295.

KNAPWEED, 190; Black Knapweed, 191.

LADY'S MANTLE, 53.
Lady's Smock, 90.
Larkspur, 75.
Leek, Wild, 332.
Lent Lily, 319.
Lily of the Valley, 329, 330.
Lily, Purple Martagon, 324, 325; White Lily, 325.
Lobelia, 207.
Loosestrife, Purple, 223; Woodland Loosestrife, 224; Yellow Loosestrife, 223.
Lords-and-Ladies, 341; a prison for flies, 343.
Lotus, 146.
Lousewort, 261.
Lucerne, 151.

MADDER, Field, 171, 173, 176.
Mallows, 118; Common Mallow, 119, 120; Dwarf Mallow, 119, 120; Hairy Marsh Mallow, 119; Common Marsh Mallow, 119.
Marjoram, 275.
Marigold, Corn, 184; Marsh Marigold, 67.
Marram or Sea Reed, 347.
Marsh Marigold, 67.
Marsh Trefoil, 234.
May, 37.
Mayweed, Scentless, 184; Stinking Mayweed, 184.
Meadow Rues, 71; Small Meadow Rue, 72; Yellow Meadow Rue, 71.
Meadow Sage, 276, 277.
Meadow Sweet, 53.
Meadow Vetchling, 139, 140.

Medick, Purple, 151; Spotted Medick, 152.
Medlar, 32.
Melilot, 152.
Menziesia, 217.
Mercury, 287, 290.
Milfoil, 183.
Milkwort, Sea, 224, 225.
Mint, 273; Corn Mint, 274; Water Mint, 275.
Moneywort, 223.
Monkey-flower, 253.
Monkshood, 75.
Mountain Ash, 34.
Mouse-ear, Broad-leaved, 111; Erect Mouse-ear, 111; Field Mouse-ear, 111; Little Mouse-ear, 111.
Mouse-tail, 72.
Mullein, 246, 248.
Mustard, Garlic, 90; Hedge Mustard, 90.

NETTLE, Hemp, 280; Henbit Dead Nettle, 283; Red Dead Nettle, 283; White Dead Nettle, 281; Stinging Nettles, 291; Great Stinging Nettle, 292; Roman Stinging Nettle, 293; Small Stinging Nettle, 293.

OAK, 295.
Oat-grass, 348.
Orchis, Bee, 302; Bird's-nest Orchis, 305; Bog Orchis, 305; Butterfly Orchis, 304; Coral-root Orchis, 305; Dwarf Orchis, 301; Early Purple Orchis, 298, 299, 300; Fly Orchis, 303; Fragrant Orchis, 304; Green-winged Orchis, 301; Helleborine, 308; Lady's Slipper, 309, 310; Man Orchis, 302; Marsh Orchis, 301; Musk Orchis, 304; Pyramidal Orchis, 302; Spotted Orchis, 298; Twayblade, 306; White Helleborine, 309.
Ox-eye Daisy, 184.
Ox-lip, 221.

PANSY, Wild, 100.
Parsley, 169.
Parsnip, Cow, 163.
Pasque-flower, 70.
Pea, Narrow-leaved Everlasting, 144; stamens and pistil of Pea, 142; Pea-pods, 142.
Pear, Wild, 36.
Pearlwort, 116.
Pellitory, 291.
Pennywort, Marsh, 167.
Pilewort, 57, 58.
Pimpernel, 225; as a weather indicator, 226; Bog Pimpernel, 226; Yellow Pimpernel, 224.
Pine, 295.
Pink, Maiden, 109.
Plum, Wild, 50.
Poplar, 295.
Poppy, Common, 82; Long Prickly-headed Poppy, 81; Long Smooth-headed Poppy, 82; Round Rough-headed Poppy, 82; Yellow-horned Poppy, 79; Welsh Poppy, 81.
Prickly Comfrey, 238.
Primrose, 218, 219; two forms of flower, 220; Bird's-eye Primrose, 222; Scottish Primrose, 222.
Purple Medick, 151.

RAGGED ROBIN, 106.
Ragwort, Common, 187; Water Ragwort, 188.
Rampion, 205; Round-headed Rampion, 206; Spiked Rampion, 206.
Rape, 91.
Raspberry, Wild, 47.
Reed, 347.
Rest-harrow, 147.
Roast-beef Plant, 317.
Rocket, London, 90.
Romulea, 312.
Rose, Dog, 24, 26; parts of Rose-flowers, 28, 29; Rose-hips, 31; Rose-leaves, 25.
Rowan, 34, 36.
Rushes, 339; Wood Rush, 340; Degenerate Lilies, 340.

SAFFRON, Meadow, 336.
Sage, Meadow, 276, 277; its remarkable adaptation to bee-form, 278; Wood Sage, 284.
Sain-foin, 145.
St. James'-wort, 187.
Sallow, 295.
Sandwort, 116; Sandwort Spurrey, 116.
Sanicle, 168.
Scabious, Devil's-bit, 200; Field Scabious, 200; Small Scabious, 200.
Scorpion Grass, 235.
Sea Holly, 167.
Sea Milkwort, 224.
Sedge, 341.
Self-heal, 280.
Service, 34; Fowler's Service, 38; Wild Service, 36.
Sheep-bit, 206.

Shepherd's Purse, 91, 92.
Silver-weed, 41.
Sloe, 50.
Snake's-head, 323, 335.
Snap-weed, 135.
Sneezewort, 185.
Snowdrop, 318, 319.
Snowflakes, Spring and Summer, 322.
Solomon's Seal, 329, 331.
Sorrel, Wood, 132.
Spearworts, Greater and Lesser, 65.
Speedwell, Germander, 248.
Spurge, Cypress, 286, 287; Irish Spurge, 289; Petty Spurge, 289; Sun Spurge, 288; Wood Spurge, 289.
Squill, Autumnal, 333; Vernal Squill, 333.
Star of Bethlehem, Spiked and Yellow, 335.
Stitchwort, Bog, 112; Greater Stitchwort, 112; Lesser Stitchwort, 112; Marsh Stitchwort, 112.
Stock, Great Sea, 88; Hoary, 88.
Stork's-bill, Hemlock-leaved, 130; Musky Stork's-bill, 130; Sea Stork's-bill, 130.
Strawberry, Barren, 41; Wild Strawberry, 41, 43.
Strawberry-tree, 211.
Sundews, 158.
Swede, 91.
Sweet Flag, 344.
Sweet William, 104.

TANSY, 186.
Teasel, 199; its insect-trap basins, 201.

Thistles, 175; Carline Thistle, 189; Creeping Plume Thistle, 194; Dwarf Plume Thistle, 194; Holy Milk Thistle, 195; Musk Thistle, 192; Spear Plume Thistle, 194; Welted Plume Thistle, 193; Woolly-headed Thistle, 195.
Thyme, Wild, 276.
Toadflax, Ivy-leaved, 263; Yellow Toadflax, 265.
Toothwort, 262.
Tormentil, 38.
Touch-me-not, 134, 135.
Trefoil, Bird's-foot, 146; Subterranean Trefoil, 156.
Tulip, 323, 335; Alpine Tulip, 336.
Turnip, 91.

VENUS' LOOKING-GLASS, 205.
Vetch, Bitter, 145; Grass Vetch, 143; Horseshoe Vetch, 146; Kidney Vetch, 147; Wood Vetch, 145.
Vetchling, Meadow, 139, 140; Yellow Vetchling, 143.
Violet, Dog, 100; Hairy Violet, 100; Marsh Violet, 100; Sand Violet, 100; Wood Violet, 100; Sweet Violet, 95.

WAKE ROBIN, 341.
Wallflower, 84, 85, 86.
Watercress, 89.
Water Crowfoot, 64, 65; Water Dropwort, 169; Water Figwort, 253; Water Hemlock, 169; Water Violet, 227.
Whortleberry, 211; Bog Whortleberry, 211.
Wild Celery, 169.
Willow, 295, 296.
Woad, 94.
Wood Anemone, 68; Wood Avens, 44; Woodruff, 171, 172; Wood Sage, 284; Wood Sorrel, 132.
Wormwood, 186.
Woundwort, 273, 280; Marsh Woundwort, 280.

YARROW, 185.
Yellow Archangel, 282; Yellow Rattle, 256; Yellow-wort, 233.
Yew, 295.

www.ingramcontent.com/pod-product-compliance
Lightning Source LLC
Chambersburg PA
CBHW030558300426
44111CB00009B/1020